T0146085

NAVY NETWORK DEPENDABILITY

MODELS, METRICS, AND TOOLS

ISAAC R. PORCHE III, KATHERINE COMANOR, BRADLEY WILSON,
MATTHEW J. SCHNEIDER, JUAN MONTELIBANO, JEFF ROTHENBERG

Prepared for the United States Navy

 NATIONAL DEFENSE RESEARCH INSTITUTE

The research described in this report was prepared for the United States Navy. The research was conducted in the RAND National Defense Research Institute, a federally funded research and development center sponsored by the Office of the Secretary of Defense, the Joint Staff, the Unified Combatant Commands, the Department of the Navy, the Marine Corps, the defense agencies, and the defense Intelligence Community under Contract W74V8H-06-C-0002.

Library of Congress Control Number: 2010934102
ISBN: 978-0-8330-4994-0

Published 2010 by the RAND Corporation
1776 Main Street, P.O. Box 2138, Santa Monica, CA 90407-2138
1200 South Hayes Street, Arlington, VA 22202-5050
4570 Fifth Avenue, Suite 600, Pittsburgh, PA 15213-2665
RAND URL: http://www.rand.org/
To order RAND documents or to obtain additional information, contact
Distribution Services: Telephone: (310) 451-7002;
Fax: (310) 451-6915; Email: order@rand.org

Preface

The Navy and the Department of Defense (DoD) are increasingly dependent on networks and associated net-centric operations to conduct military missions. As a result, a vital goal is to establish and maintain dependable networks for ship and multiship (e.g., strike group) networks. An essential step in maintaining the dependability of any networked system is the ability to understand and measure the network's dependability. The term *network dependability* is broad. It is determined, in part, by the *availability* and *reliability* of information technology (IT) systems and the functions these systems provide to the user. For the Navy, qualitative standards for network dependability include (1) the ability of the Navy's IT systems to experience failures or systematic attacks without impacting users and operations, and (2) achievement of consistent behavior and predictable performance from any access point.

The RAND National Defense Research Institute (NDRI) was asked to develop an analytical framework to evaluate C4I (command, control, communications, computers, and intelligence) network dependability. This requires an understanding of the availability and reliability of the network and its supporting systems, subsystems, components, and subcomponents. In addition, RAND was asked to improve upon an existing tool—initially developed by the Space and Naval Warfare Systems Command (SPAWAR)—to help better evaluate network dependability. This report documents these efforts.

This research was sponsored by the U.S. Navy and conducted within the Acquisition and Technology Policy (ATP) Center of the RAND National Defense Research Institute, a federally funded research and development center sponsored by the Office of the Secretary of Defense, the Joint Staff, the Unified Combatant Commands, the Department of the Navy, the Marine Corps, the defense agencies, and the defense Intelligence Community. Questions and comments about this research are welcome and should be directed to the program director of ATP, Philip Antón (anton@rand.org), or the principal investigator, Isaac Porche (porche@rand.org).

For more information on RAND's Acquisition and Technology Policy Center, contact the Director, Philip Antón. He can be reached by email at atpc-director@rand.org; by phone at 310-393-0411, extension 7798; or by mail at the RAND Corporation, 1776 Main Street, P.O. Box 2138, Santa Monica, California 90407-2138. More information about RAND is available at www.rand.org.

Contents

Figures

Tables

Summary

The Problem

The Navy and DoD are increasingly dependent on networks and associated net-centric operations to conduct military missions. As a result, a vital goal is to establish and maintain dependable networks for ship and multiship (e.g., strike group) networks. An essential step in maintaining the dependability of any networked system is the ability to understand and measure the network's dependability. The problem is that the Navy does not do this well. Existing metrics, as we will discuss, are insufficient and inaccurate; for example, they are not always indicative of user experiences.

The term *network dependability* is broad. It is determined, in part, by the *availability* and *reliability* of IT systems and the functions these systems provide to the user. For the Navy, qualitative standards for network dependability include (1) the ability of the Navy's IT systems to experience failures or systematic attacks without impacting users and operations, and (2) achievement of consistent behavior and predictable performance from any access point.

The complexity of shipboard networks and the many factors[1] that affect the dependability of a network include

- hardware
- software applications/services
- environmental considerations (e.g., mission dynamics)
- network operations
- user (human) error
- network design and (human) process shortfalls.

The above list was compiled from many sources, including academia and data from the fleet (Conwell and Kolackovsky, 2009). For example, the literature (Tokuno and Yamada, 2008; Gray and Siewiorek, 1991) reports that software faults and human errors are significant root causes of outages in computer systems.

The dynamics of a mission are also a factor affecting network dependability. Specifically, geographic and geometric effects (e.g., line of sight for radio frequency trans-

---

[1] Subfactors exist: For example, training levels and sufficiency of training and operating manuals can impact human errors.

missions, ships' motions) may cause many intermittent line-of-sight blockages from ships' infrastructures. The relative position of each ship can also affect interference. These effects are particularly important with regard to satellite communications.

Prior to our study, an effort to develop a tool to understand network availability and reliability for specific missions was undertaken by SPAWAR. This report builds upon that work by applying new insights to produce improved modeling of dependability overall and availability and reliability in particular. We did this by making changes to the initial SPAWAR tool to improve it. Among the changes: the incorporation of new and more recent data sources, use of more detailed architectural considerations, and the inclusion of uncertainty for most of the availability and reliability measurements.

The overall objectives of this research were twofold: to better understand the shortfalls in developing an accurate awareness of network dependability (by the Program Executive Office [PEO] C4I) and to apply some of the resulting lessons learned to the enhancement of an existing SPAWAR tool. We discuss some of the drivers and causes for the inadequacy in current high-level C4I readiness reporting systems.

Measuring Dependability Today

Again, *network dependability* is an overarching term that is accounted for in part by the availability and reliability of networks and IT. The term *network dependability* can be used to describe (1) the ability of the Navy's IT systems to experience failures or systematic attacks without impacting users and operations, and (2) achievement of consistent behavior and predictable performance from any network access point. There is not a single universally accepted measure for dependability when it comes to IT systems and/or networks.

Nonetheless, both qualitative and quantitative standards are in use by the Navy. *Operational availability* (Ao) is quantified in a standardized way in many requirements documents. Operational availability is defined by OPNAV Instruction 3000.12A as follows:

$$Ao = Uptime/(Uptime + Downtime).$$

In earlier SPAWAR work, Larish and Ziegler (2008a) applied the Ao metric to strings of equipment that are necessary to accomplish a mission; they also developed the concept of *end-to-end mission Ao* (hereafter referred to as *mission availability*) as a metric for the availability of all the equipment threads necessary to support a given mission.

However, the Navy's traditional definition of Ao is too general when applied to networks and fails to capture the nuances of network operations. The fact that the Ao metric is based on hard definitions of *uptime* and *downtime* and does not account for "gray areas" of network performance, such as network congestion and degradation, is a shortfall. When networks are congested, for example, some users may perceive the network as being fully functional, while others users (and other metrics) may not. Partly as a result of the way that Ao is defined, there is a gap between what network users experience in terms of dependability and what existing fleet operational availability data suggest. This report attempts to explain why this gap exists and suggests measures that are needed to bridge the gap.

Measuring IT Dependability: User-Perceived Service Availability

A better framework for measuring network dependability should consider users' perceptions of the dependability of the specific services that the network provides.[2] The framework we propose incorporates the following:

- the types of services available to shipboard users
- the volume of user requests for these services
- the availability of these individual services
- the impact of these services on various missions as well as the relative importance of these missions (not assessed in this work).

We propose a framework for a new service-based availability metric: *user-perceived service availability*. Our goal is to re-orient the model to the user perspective by modeling availability of services used by individuals to accomplish a mission. This was motivated by the work of Coehlo et al. (2003). The framework proposed follows three ideas: (1) model the service the user is trying to invoke (by factoring the availabilities of the hardware, software, and human interaction involved), (2) weight the number of users who can or do invoke the service, and (3) weight the mission impact of each service (not assessed in this study). While these three ideas are all discussed in our proposed framework, they are not all implemented in our network availability modeling tool. Future work is needed to further enhance the tool so as to fully incorporate all of the ideas in the proposed framework.

Drivers of Dependability, Availability, and Reliability

Reports suggest that hardware (equipment) failures are not the sole root cause of failures of IT systems. Within PEO C4I, a CASREP (Casualty Report) study was recently completed using 169 CASREPs recorded aboard carriers (Conwell and Kolackovsky, 2009). Analysis of these CASREPs indicated that most of the problems fell into nine

---

[2]   For this report, a *service* is a discrete IT-based function/capability, such as a "chat" application, that meets an end-user need.

categories, which can be considered "first-order" root causes. Specifically, Conwell and Kolackovsky binned these root causes as follows:

1. Hardware (37 percent, or 63 out of 169)
2. Training (16 percent, or 27 out of 169)
3. Integrated Logistics Support (ILS) (12 percent, or 20 out of 169)[3]
4. Design (11 percent, or 19 out of 169)
5. Configuration Management (9 percent, or 15 out of 169)
6. Settings (4 percent, or 7 out of 169)
7. Software (4 percent, or 7 out of 169)
8. System Operational Verification Testing (SOVT) (3 percent, or 5 out of 169)
9. Other (4 percent, or 7 out of 169).

These causes are shown in the pie chart in Figure S.1. What is clear is that nearly two-thirds of the CASREPs do not directly involve hardware/equipment failures.

If we create a meta-category called "human and human-defined processes" and include training, design, SOVT, and ILS within it, then this meta-category accounts for 42 percent of CASREPs using the study data. If we include software, configuration management, and settings into a more general "software" category, then these causes account for 16 percent of CASREPs. It is important to note as well that configuration management and settings could have root causes of training, inattention to detail,

Figure S.1
Root Causes of CASREPs

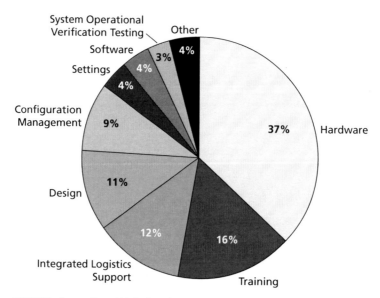

SOURCE: Conwell and Kolackovsky, 2009.
RAND *MG1003-S.1*

---

[3]   The ILS category includes parts not being onboard and technical manuals not being up to date or available.

or other human process factors. Hence, the 42 percent statistic from this study data is really a lower bound on the percentage of CASREPs that result from human and human-defined processes, as Conwell and Kolackovsky's first-order root causes do not necessarily map perfectly into human- and nonhuman-driven error categories.

Conwell and Kolackovsky's observations are consistent with other literature that finds that more than 50 percent of failures in networked IT environments are caused by human error (Snaith, 2007; Kuhn, 1997; "Human Error Is the Primary Cause of UK Network Downtime," 2003). Additional studies flag software: Hou and Okogbaa (2000) gathered industry data points and concluded that software is a major cause of failures in networks rather than just hardware for systems that involve both.

In summary, hardware equipment failures alone cannot account for gaps in network dependability. In addition to hardware/equipment failures, two significant drivers of dependability that cannot be ignored are (1) flawed human processes and human error, and (2) software problems. It is important to note that some academic publications dedicated to the topic of human error in network reliability analysis focus on the issue of interface design. Although Conwell and Kolackovsky (2009) did not identify interface design as a root-cause category in their CASREP assessment, interface design could be a significant contributing factor to a number of the CASREP causes. Improved interfaces must be considered as part of the larger solution set for improving Navy shipboard network dependability.

What We Can Use Now for Modeling Availability

RAND evaluated an existing spreadsheet that was designed by SPAWAR to model network availability for a specific antisubmarine warfare (ASW) mission. RAND built upon this model by

- incrementally expanding the resolution of the model to include some system components
- using nondeterministic values for mean time between failure (MTBF) and mean down time (MDT) from fleet data (Naval Surface Warfare Center–Corona [NSWC-Corona]) instead of fixed required values in requirements documents
- parameterizing values when data (MTBF, MDT) are missing, unavailable, or assumed (MDT was broken down into the components of mean time to repair [MTTR] and mean logistic delay time [MLDT])
- enabling Monte Carlo–based sensitivity analysis on a per-component or -system basis
- making the model updatable and user-friendly: allowing for constant changes in data based on field measurements of Ao
- making the model user-friendly: adding a front-end (VBASIC) graphical user interface (GUI) to allow users (e.g., decisionmakers) to modify assumptions and certain values/factors

- developing exemplar models for both SameTime chat and common operational picture (COP) services to better evaluate user perceived service availability.

Analysis of component availability using the newly modified tool (or "new tool") was done and is described in this report.

Exemplar Analysis Results

Antisubmarine Warfare Mission

An ASW mission was utilized in the original SPAWAR effort that analyzed mission threads to identify strings of equipment whose availabilities could impact the mission. Details of this effort are described by Larish and Ziegler (2008a). We briefly describe the ASW mission using the operational view illustrated in Figure S.2.

Figure S.2
Operational View of the Antisubmarine Warfare Mission

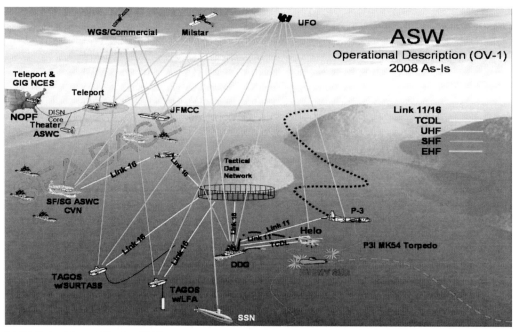

SOURCE: Larish and Ziegler, 2008a.
NOTES: ASWC = Antisubmarine Warfare Commander; EHF = extremely high frequency; GIG = Global Information Grid; JFMCC = Joint Force Maritime Component Commander; LFA = Low Frequency Active; NECS = Net-Centric Enterprise Services; NOPF = Naval Ocean Processing Facility; SHF = super high-frequency; SURTASS = Surveillance Towed Array Sensor System; TAGOS = tactical auxiliary general ocean surveillance; TCDL = Tactical Common Data Link; UHF = ultra high frequency; UFO = ultra-high-frequency follow-on; WGS = Wideband Gapfiller Satellite.

Figure S.3 illustrates three equipment strings used in the ASW mission: an "IP (Internet Protocol) network" string that enables ship-to-shore data exchange, a secure voice string that enables secure voice communications, and a "tactical datalink" string that enables other data exchanges among aircraft, ship, and shore. These equipment strings enable exchanges across the network that are required for the ASW mission.

Each of these equipment strings consists of many individual components. For example, Figure S.4 shows a component-level view of the secure voice equipment string.

Exemplar Sensitivity Analysis

We used the new tool to determine which components of the ASW mission network have the greatest impact on overall mission availability, as well as which components have the greatest impact on the availability of specific equipment strings within the network.

We found that the Digital Modular Radio, or DMR (shown as a component on the DDG in Figure S.4), is the most sensitive component when considering the entire ASW mission Ao. Specifically, a one-standard-deviation increase in the DMR's mean down time results in a 2-percent decrease in the mean value of the mission Ao.

Figure S.3
Equipment Strings Used in the ASW Mission

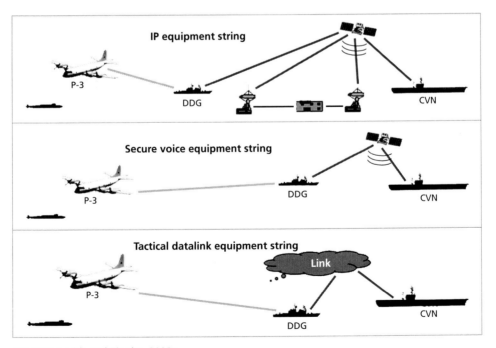

SOURCE: Larish and Ziegler, 2008a.
RAND *MG1003-S.3*

Figure S.4
Secure Voice Equipment String: Component-Level View

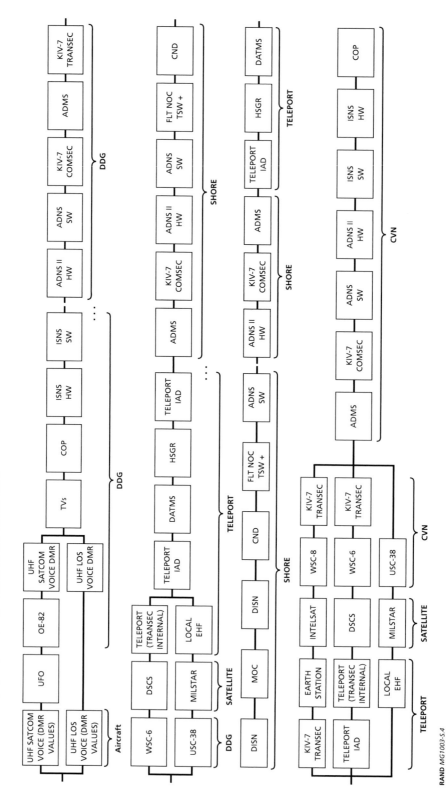

RAND MG1003-5.4

We also analyzed equipment strings in isolation. For example, for the link equipment string, we found that Common Data Link Monitoring System (CDLMS, AN/UYQ-86) is the most vital component in terms of the availability of the equipment string. Specifically, a one-standard-deviation increase in the mean down time of the CDLMS results in a 4-percent decrease in the link equipment string's overall availability.

Comparing the New Tool with the Old Tool

We summarize here the results that compare the original SPAWAR 5.1.1 spreadsheet model (the old tool) with RAND's modifications to it (the new tool) for Ao values in the ASW mission's equipment strings. A major difference in results between the new model and the old one is that the new one is far less optimistic about component availability. This is due to the inclusion of real-world data and uncertainty. The major differences between the two efforts are enumerated as follows:

- The old model (SPAWAR 5.1.1 Pilot Effort):
  - used requirements' threshold and objective values for system and component performance specifications (MTBF and MDT)
  - employed a deterministic model (e.g., factored in no uncertainty) and thus generated a single value for equipment string and overall mission availability.
- The new model (RAND's Modified Model):
  - uses historical data (where available) instead of purely data from requirements documents
  - employs a stochastic modeling approach:
    o fits historical data to probability distributions to describe system and component performance specifications (MTBF and MDT)
    o generates distributions describing probability of a given availability for equipment strings and for the overall mission.

RAND's modified tool adds additional functionality to the old model as well. Specifically, the new tool performs sensitivity analysis of the systems and components in the model to determine their relative importance on the individual equipment strings and overall mission availability. For example, the new tool allows the user to isolate particular network segments, such as the Automated Digital Network System (ADNS) and the Integrated Shipboard Network System (ISNS), and to perform separate analyses on that portion of the network.

Figure S.5 shows a large range of potential availabilities. This is because the new tool accounts for uncertainty in measures that drive availability. Ninety percent confidence intervals are shown on the plots in darker blue. Means are shown as white dots.

Figure S.5
Comparison of Mission Availability: Old Versus New Model

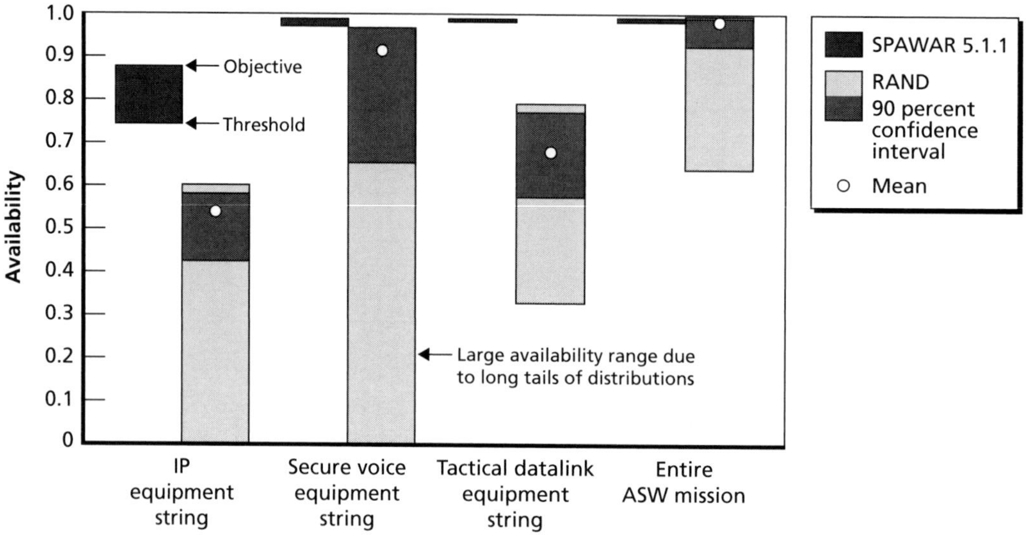

RAND *MG1003-S.5*

The Way Ahead

The fundamental issue that motivated this research effort was communicated to us by the sponsor (PEO C4I) as follows. There is a need to understand why perceptions about the dependability (e.g., availability) of networks from users' (e.g., sailors) perspectives sometimes differ from the availability data acquired from the "usual" sources (e.g., NSWC-Corona Reliability, Maintainability, and Availability [RMA] data). We did not attempt to substantiate users' perception of a lack of network availability by directly surveying them. But the correctness of this perception seems likely from our investigations, which identified the many factors that impact network dependability that are not accounted for well today. The way ahead for this research and the Navy is to consider ways to fold the additional factors into today's assessment process for network dependability. We enumerate some of the ways to do this as follows.

Fuse More Data Sources to Model Dependability of Networks

During the course of this study, we encountered numerous sources of availability and reliability data. NSWC-Corona is the most visible source of RMA data, and, as a result, they served as our primary source for component data for use in our models. However, in addition to NSWC-Corona RMA data, which mainly documents hardware failures, there are other data sources that contain Remedy Trouble Ticket System (RTTS) data, which could be used to take into account user-reported failures.

New methods should be pursued to take into account such user-reported failures as recorded in TTS data, which include

- Remedy (for ADNS and ISNS)
- Fleet Systems Engineering Team (FSET)
- Information Technology Readiness Review (ITRR)
- Board of Inspection and Survey (INSURV)
- Deploying Group Systems Integration Testing (DGSIT).

In this study, we were able to study ISNS and ADNS trouble-ticket data from Remedy to analyze human-error impacts on availability. According to a Gartner survey (2007), this is done in the commercial world by more than a few organizations. Future models could use this approach to develop predictions of the human impact on network dependability.

In summary, there are other data sources (beyond availability data from NSWC-Corona) that can provide valuable insight into network dependability. However, due to their varying formats, they did not lend themselves to a standardized data-extraction process. The Navy could standardize trouble ticket system data so that they have uniform reliability metrics. However, there is no central repository to consolidate all these data, making analysis difficult.

Rely on User-Perceived Service and Mission Availability Metrics

We believe that user-perceived service availability is an important metric for Navy networks. User-perceived service availability can be defined as the number of correct (e.g., successful) service invocations requested by a particular user for a given number of total service invocations requested by that user, for a given time interval. The Navy can leverage current efforts by the Information Technology Readiness Review to develop data on the success and failure of a service invocation.

Modeling the availability of a particular service is easier to conceive than to calculate: An understanding (e.g., diagrams) of all the hardware and software involved—as well as the human interaction that takes place—will be needed to account for what composes the service (e.g., chat). Such an effort could prove worthwhile because it could yield a better assessment of IT and network dependability, e.g., one that matches user sentiment of dependability.

Further Enhance Existing Tools

As suggested by the PMW 750 CVN C4I CASREP study report (Conwell and Kolackovsky, 2009), PEO C4I has a need to establish a C4I performance dashboard using NSWC-Corona data.[4] The framework and tool described in this research can con-

---

[4] PMW 750 is an organization within PEO C4I that serves as the integrator for C4I systems.

tribute to that goal. Toward this goal, RAND recommends that the newly developed network availability modeling tool described herein be made web-based and upgraded to automatically incorporate the latest historical data from NSWC-Corona as they are made available. It should also be made to fuse other data sources, as described above, and to incorporate mission impacts to allow more-relevant sensitivity analysis.

Create a Single Accessible Portal for Network Diagrams

PEO C4I should consider a way to facilitate a holistic view of all afloat and shore networks. A major challenge to this study was to gather and consolidate disparate network diagrams to come up with a comprehensive end-to-end architecture. SPAWAR has made initial inroads in this task with the development of a website (not available to the general public) intended specifically for in-service engineering agents (ISEAs), which contains network diagrams for afloat platforms. The addition of shore network diagrams to this website would greatly facilitate the reliability study of additional network services.

Acknowledgments

This project was sponsored by Chris Miller, Program Executive Officer Command, Control, Computer, Communications, and Intelligence (PEO C4I). We were guided by the project monitor, Captain Joe Beel, Deputy Program Manager of PMW 160 within PEO C4I. Joe Frankwich worked hard to get data to us, as did Mitch Fischer at NSWC-Corona. We had helpful discussions with many personnel from PMW 160. This includes Brian Miller, Chuck Tristani, and Tom Chaudoin, who shared key reports that were used for this analysis. Brian Miller served as the central point of contact throughout the network data–gathering portion of this study and coordinated a large majority of our meetings at SPAWAR. We are also grateful to Diego Martinez, Dai Nguyen, and Kyle Wheatcroft at PMW 790, and to CDR Pat Mack, Steve Roa, and Rich Kadel at PMW 150. Roderick Zerkle, David Klich, and Eric Otte lent their technical expertise to our team. Hal Leupp, John Datto, and Janet Carr provided a good portion of data and insight to help complete the Human Error portion of this study. We collaborated with some remote in-service engineering agents (ISEA) and subject-matter experts whom we did not have the pleasure of meeting in person, but who still contributed a good deal to this study, including Daryl Ching and Rob Sotelo. Many others in PEO C4I and SPAWAR provided feedback and assistance with data collection, including Charlie Suggs, Jack Cabana, and Aaron Whitaker. We would also like to thank John Birkler, John Schank, John Hung, Mike H. Davis, Sean Zion, and Rob Wolborwsky for their overall guidance. We thank our reviewers, Paul Dreyer of RAND and Steve Sudkamp of the Johns Hopkins University Applied Physics Laboratory. Michelle McMullen and Sarah Hauer provided assistance in preparing this document. Finally, Bryan Larish of SPAWAR and Michael Ziegler (and his colleagues at Systems Technology Forum, Ltd.) spent countless hours explaining their initial tool development efforts. We are grateful.

Abbreviations

ADMS	Automated Digital Multiplexing System
ADNS	Automated Digital Network System
ASW	antisubmarine warfare
Ao	operational availability
C4I	command, control, communications, computers, and intelligence
CANES	Consolidated Afloat Networks and Enterprise Services
CASREP	Casualty Report
CDLMS	Common Data Link Monitoring System
CENTRIXS	Combined Enterprise Regional Information System
CND	computer network defense
COP	common operational picture
COTS	commercial, off the shelf
CST	COP Synchronization Tool
CTI	category/type/item
CVN	aircraft carrier (nuclear)
DDG	destroyer
DGSIT	Deploying Group Systems Integration Testing
DMR	Digital Modular Radio
DNS	Domain Name System
DOA	Determination of Availability
DoD	U.S. Department of Defense
E2E	end-to-end

FSET	Fleet Systems Engineering Team
GCCS	Global Command and Control System
GUI	graphical user interface
HMI	human-machine interaction
HP	Hewlett-Packard
HW	hardware
ILS	Integrated Logistics Support
INSURV	Board of Inspection and Survey
IP	Internet protocol
ISEA	in-service engineering agent
ISNS	Integrated Shipboard Network System
ISO	International Organization for Standardization
IT	information technology
ITRR	Information Technology Readiness Review
LOS	line of sight
MDT	mean down time
MIDS	Multi-functional Information Distribution System
MLDT	mean logistic delay time
MRT	mean restoral time
MTBF	mean time between failure
MTTR	mean time to repair
NOC	network operations center
NSWC	Naval Surface Warfare Center
OPNAV	Naval Operations
PEO	program executive office
PMW 750	Carrier and Air C4I Integration Program Office
POR	Programs of Record
RBD	reliability block diagrams
RMA	Reliability, Maintainability, and Availability
RMC	Regional Maintenance Centers

RTTS	Remedy Trouble Ticket System
SATCOM	satellite communications
SOA	service-oriented architecture
SOVT	System Operational Verification Testing
SPAWAR	Space and Naval Warfare Systems Command
UHF	ultra high frequency
URL	Uniform Resource Locator
VTC	video teleconference
XML	Extensible Markup Language

CHAPTER ONE

Introduction

Background

The Navy and the Department of Defense (DoD) are increasingly reliant on networks and associated net-centric operations to conduct military missions. This puts a premium on establishing and maintaining dependable individual ship and multiship (e.g., strike group) networks. An essential step in maintaining the dependability of any networked system is the ability to understand and measure the network's dependability.

Objectives of This Study

The Program Executive Office (PEO) Command, Control, Communications, Computers, and Intelligence (C4I) tasked the RAND National Defense Research Institute (NDRI) with developing an analytical framework to evaluate C4I network dependability and its contribution to the operational capabilities of individual ships and multiship formations. To this end, we proposed a two-phased approach:

- Phase 1: Model network dependability.
- Phase 2: Quantify the operational importance of C4I network dependability to the operation capabilities of individual ships and multiship formations.

This report describes the Phase 1 effort. The subtasks under Phase 1 were as follows:

- Task 1.1: Study the architecture that composes the network/infostructure.
- Task 1.2: Define network dependability and how it should be measured for Navy use.
- Task 1.3: Use given data on dependability to further determine factors driving dependability.
- Task 1.4: Associate factors that drive network dependability with parts of the network architecture/infostructure.
- Task 1.5: Create a dependability model with material and nonmaterial factors as variables.

Approach and Methodology

Our primary goal was to develop a tool that would allow users to determine the impact of modified component performance on the overall availability of a given network service, or on the overall availability of all the network services necessary to support a given mission. To this end, we developed a network availability modeling tool by further expanding on an existing Excel spreadsheet model developed by Space and Naval Warfare Systems Command (SPAWAR). We modified the SPAWAR tool with a Monte Carlo add-on package, @RISK, and an easy-to-use graphical user interface (GUI). This newly developed tool incorporates stochastic models of system and component performance. Using Monte Carlo simulations, our tool allows the user to

1.   visualize a range in potential mission or service availabilities, as shown in the availability histogram that results from the Monte Carlo runs
2.   perform sensitivity analysis that captures the degree to which each of the components and systems in the mission or service in question impact the overall operational availability (Ao).

With this tool in hand, we set out to define a measure of network reliability that differed from that presented in the SPAWAR 5.1.1 report (Larish and Ziegler, 2008a). We re-oriented the model to capture user-perceived service availability by simulating the availability of specific services rather than focusing on a string of hardware components.

The driving force behind this new approach was the disparity among the feedback provided to PEO C4I from the engineers and the users in the field: Field data might suggest high availability, whereas users would consistently complain about network outages. This is because classical metrics (i.e., hardware string measurements) employed by engineers yielded artificially inflated availability values, some even reaching the goal of "five nines" operational availability (99.999 percent). However, users would often complain that networks were anything but reliable, with some platforms reporting failures up to "half the time we were underway." The reason for this discrepancy is that the traditional way of measuring availability focuses on specific pieces of *equipment*, whereas users focus on whether they are able to use a *service*, the availability of which depends not just on equipment but also on the users' actions, software, and environmental factors. By re-orienting the model to a user perspective, our goal is to more realistically simulate user experience.

PMW 160 provided RAND with data from its Remedy Trouble Ticket System (RTTS) database to demonstrate the underlying unreliability of afloat networks. RTTS data serve to better "tune" the model to make it more accurate. We illustrate our approach by modeling two services—SameTime chat and common operational picture (COP).

To create accurate reliability block diagrams (RBDs) for selected services, we interviewed subject-matter experts and in-service engineering agents (ISEAs) for the

Integrated Shipboard Network System (ISNS), Automated Digital Network System (ADNS), network operations center (NOC), satellite communications (SATCOM), and Global Command and Control System (GCCS) architectures. We also worked closely with Naval Surface Warfare Center–Corona (NSWC-Corona) to gather historical data for the antisubmarine warfare (ASW) mission, SameTime chat service, and COP service. RAND engineers ensured that subject-matter experts and ISEAs validated the resulting network diagrams and RBDs before moving forward with the modeling efforts.

Antisubmarine Warfare Mission

This report makes reference to an ASW mission used to develop operational availability calculations. This section provides a sufficient level of detail to assist the reader in understanding how this mission thread, illustrated in Figure 1.1, was analyzed. Work

Figure 1.1
Operational View of the Antisubmarine Warfare Mission

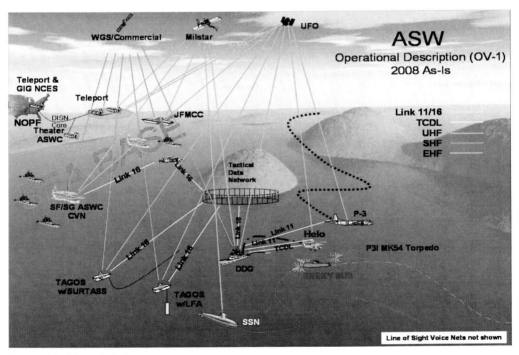

SOURCE: Larish and Ziegler, 2008a.
NOTES: ASWC = Antisubmarine Warfare Commander; EHF = extremely high frequency; GIG = Global Information Grid; JFMCC = Joint Force Maritime Component Commander; LFA = Low Frequency Active; NECS = Net-Centric Enterprise Services; NOPF = Naval Ocean Processing Facility; SHF = super high-frequency; SURTASS = Surveillance Towed Array Sensor System; TAGOS = tactical auxiliary general ocean surveillance; TCDL = Tactical Common Data Link; UHF = ultra high frequency; UFO = ultra-high-frequency follow-on; WGS = Wideband Gapfiller Satellite.
RAND *MG1003-1.1*

by Larish and Ziegler (2008a) decomposed this mission thread into "three parallel equipment strings" that are used repeatedly in the ASW mission:

- **Internet Protocol (IP) Network Equipment String:** This string is associated with network and communications transport, servers, and the loaded software applications. This string includes an aircraft (e.g., P-3)-to-DDG line-of-sight secure voice communication path at the tactical edge (see Figure 1.2).
- **Secure Voice Equipment String:** This string is composed of over-line-of-sight and satellite communications between aircraft, submarines, ships, and shore nodes (see Figure 1.3).
- **Tactical Datalink Equipment String:** This string connects aircraft, ship, and shore command center nodes (see Figure 1.4).

Each of these equipment strings consists of many individual components. Note that availability data affects the model calculations, and data collected for one version of a component may be different from another version. Figure 1.5 shows a component-level view of the secure voice equipment string.

Organization of This Report

The main body of this report is in three parts. The first part consists of Chapter Two and Chapter Three. These chapters provide a review of the literature on definitions and metrics for dependability, availability, and reliability. Specifically, Chapter Two discusses measures of dependability, and Chapter Three discusses factors that drive reliability.

The second part reviews available data and consists of Chapter Four and Chapter Five. Chapter Four reviews the data sources for developing models of dependability in the Navy. Chapter Five provides data on specific components that make up selected network services (e.g., a model of a chat and model of a COP service).

The third part, consisting of Chapters Six and Seven, provides details on the development and refinement of the RAND network availability modeling tool and the user-perceived service availability approach developed for it. Specifically, Chapter Six describes the tool and how it was developed to calculate availability for the ASW mission. Chapter Seven presents an exemplar of the analysis that can be obtained by using the tool described in Chapter Six.

The main body of this report is followed by conclusions, recommendations, and next steps for this research effort. These are in Chapter Eight.

The appendixes contain the following: more-detailed descriptions of mathematical methods, network diagrams, examples of how data can be used to model human factors, and accompanying documents used in support of this report. Appendix A contains RAND's evaluation of the SPAWAR 5.1.1 End-to-End Operational Availability of ASW study. Appendix B reviews data collected from selected ship's network-related casualty reports and examines the human impact on network dependability. Appendix C provides additional network diagrams and detail for services modeled.

Figure 1.2
IP Network Equipment String

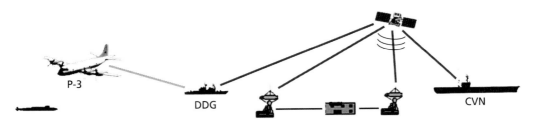

SOURCE: Larish and Ziegler, 2008a.
RAND *MG1003-1.2*

Figure 1.3:
Secure Voice Equipment String

SOURCE: Larish and Ziegler, 2008a.
RAND *MG1003-1.3*

Figure 1.4
Tactical Datalink Equipment String

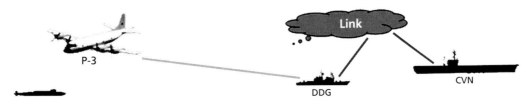

SOURCE: Larish and Ziegler, 2008a.
RAND *MG1003-1.4*

Figure 1.5
Secure Voice Equipment String: Component-Level View

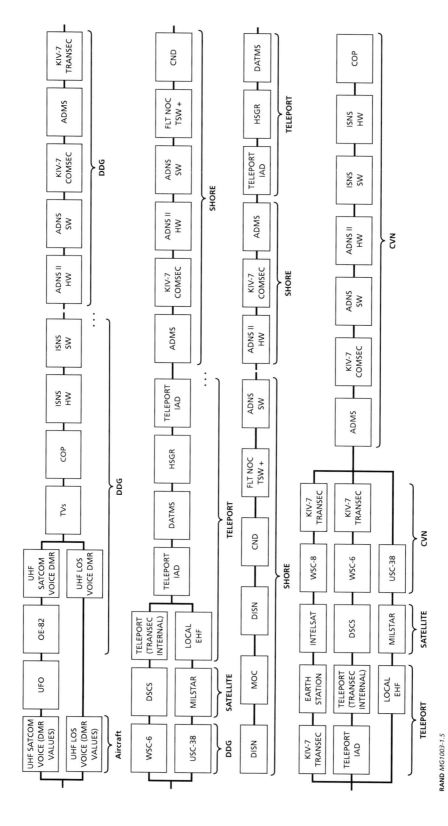

CHAPTER TWO

Measures of Dependability

This chapter provides a survey of the literature on measuring network dependability, availability, and reliability. In addition, we draw conclusions on the best variant of these to use with respect to the objectives of this research effort.

Attributes of Dependability

As the importance of network dependability, availability, and reliability has increased over the past decades, some have turned to Jean-Claude Laprie's (1992) definition of these seemingly synonymous terms for clarification, and we do as well. Laprie used the term *dependability* as an all-encompassing concept for all other descriptive terms regarding network performance. In this scheme, reliability and availability are subsets of the concept of dependability. Laprie expressed this relationship graphically, as shown in Figure 2.1.

Figure 2.1
Dependability and Its Attributes

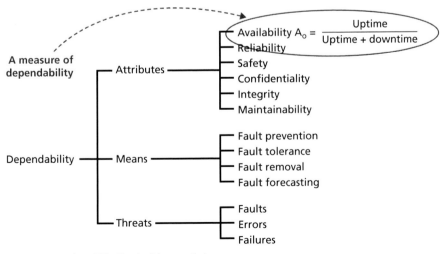

SOURCE: Laprie, 1992. Used with permission.
RAND *MG1003-2.1*

It is important to note that the terms *availability* and *reliability* are often interchangeable in a general sense. However, each term does have a distinct definition. From a strictly academic and mathematical perspective, the term *reliability* describes the probability that a service[1] will be continuously available over a given period of time, or a time interval.

Quoting Vellela (2008): "The reliability of a system . . . represents the probability that the system will perform a mission action without a mission failure within a specified mission time, represented as *t*." Hence, a system with 90 percent reliability has a 90 percent probability of operating continuously throughout the mission duration *t* without a critical failure. The reliability of a system can be expressed as $R(t) = e^{-\lambda t}$, where lambda (λ) represents the failure rate, that is, the frequency of failure occurrences over time.

Availability can be defined as the probability of a service delivering its intended function at a given point in time.[2] A related term shown in Figure 2.1 is *maintainability*, which is defined as the time it takes to restore a system to a specified condition when maintenance is performed by forward support personnel having specified skills using prescribed procedures and resources to restore the system.

Navy Definition of Operational Availability

Many Navy requirements documents (e.g., the Consolidated Afloat Networks and Enterprise Services [CANES] Capability Development Document [CDD]) specify a requirement for *operational availability* (Ao). A practical definition is that operational availability is the ability of a product to be ready for use when the customer wants to use it, or, as Vellella et al. (2008) put it, "the percentage of calendar time to which one can expect a system to work properly when it is required."

OPNAV Instruction 3000.12A provides several definitions and equations to calculate operational availability. One of them is as follows:

$$Ao = Uptime / (Uptime + Downtime).$$

---

[1] For this report, a service is a discrete IT-based function/capability, such as a "chat" application, that meets an end-user need.

[2]   Roughly, an availability of 90 percent (referred to as "one nine") for a component could imply a downtime of 36.5 days in a year for that component. An availability of 99 percent (two nines) could imply a downtime of 3.65 days in a year. Similarly, the following availabilities are paired with their implied downtimes as follows: 99.9 percent (three nines) and 8.76 hours/year, 99.99 percent (four nines) and 52 minutes/year, 99.999 percent (five nines) and 5 minutes/year, 99.9999 percent (six nines) and 31 seconds/year.

A more precise equation uses the following additional measures:

- mean time between failure (MTBF)
- mean time to repair (MTTR)
- mean logistics delay time (MLDT), which is a supportability measure.

Using these terms, Ao can be calculated as follows:

$$Ao = MTBF / (MTBF + MTTR + MLDT).$$

In earlier SPAWAR work, Larish and Ziegler (2008a) applied the Ao metric to strings of equipment that are necessary to accomplish a mission; they also developed the concept of end-to-end (E2E) mission Ao (hereafter referred to as *mission availability*) as a metric for the availability of all the equipment threads necessary to support a given mission.

Industry Measures

Industry has typically utilized a number of metrics to quantify dependability, availability, and reliability. Some of these are similar to those we have discussed, while others are more detailed and specific. We enumerate some of the metrics we found in open source briefings and in the literature:

- metrics related to customer service (call center) desks:
  - average speed to answer
  - call avoidance
  - mean time to repair for all severity levels
  - first-contact resolution
  - number of contacts per month per employee
  - channel delivery mix
  - abandonment rate
  - number of incidents caused by improper changes
- metrics related to infrastructure:
  - application response
  - network performance (bandwidth, latency)
  - voice and video (jitter, delay, packet loss)
- other, noncategorized metrics:
  - downtime, unavailability of services (minutes)
  - incident detection, response, repair, recovery, and restoration times
  - incident response times
  - number of repeat failures

- mean time between system incidents
- percentage of time that a service level agreement can be satisfied (Wang et al., 2007).

This list provides a sense of how organizations outside the Navy have attempted to measure dependability.

ISO Standards

It is important for the sake of completeness to mention that the International Organization for Standardization (ISO) has standards that are somewhat pertinent to this topic. ISO 20000 (IT Service Management)/ISO 10012:2003 (Measurement Management Systems) is somewhat relevant. However, these standards do not define specific metrics or specific processes. They simply emphasize the importance of documentation and calibration of measurement devices. They do emphasize the importance of service-level agreements (SLAs).

Service Availability

Interval service availability is the number of correct service invocations over a number of total service invocations for a given time interval. Service availability is partly a function of the hardware components of the network.

Consider a typical hardware-oriented network diagram depicting a client communicating with a server, as shown in Figure 2.2. In this "classical" approach, the availability of the service is determined by the functional chain of hardware components between the client and the server—in this case, a switch and a router.

Consider, however, if this simplified example was intended to represent a web client requesting a web page from a server. A critical part of all web communications is a Domain Name System (DNS) server. However, the DNS server rarely (if ever) falls in the serial path between a client and a server. Rather, the DNS server almost always resides in a separate network, and, for the client's request to succeed, the DNS server has to fulfill its function, namely, to resolve the requested Uniform Resource Locator

Figure 2.2
Classical Representation of Client-Server Communication

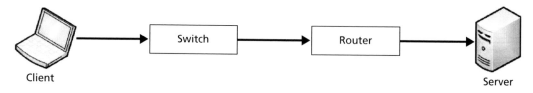

Figure 2.3
Revised Network Diagram with DNS Server

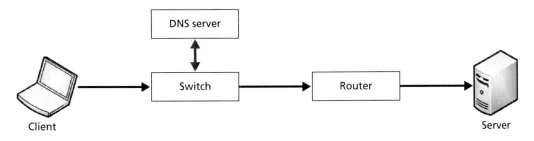

RAND *MG1003-2.3*

(URL, e.g., http://www.navy.mil) to its corresponding IP address (for www.navy.mil, the IP address is 208.19.38.89). The revised service-oriented diagram would be depicted as shown in Figure 2.3.

Figure 2.3 is a more realistic representation of the web service, because it captures the true sequence of events involved in a web transaction. The client first requests the IP address of the requested web page from a DNS server, and only once that request is fulfilled can the client continue to its ultimate transaction, which is to request a web page from the appropriate server. The classical hardware-oriented approach to network reliability fails to capture this dependency.

The idea of tracking service availability is not new and not without some concern. According to Malek et al. (2008),

> Classical analytical methods have been applied to determine service availability with mixed success, partially because many approaches failed to incorporate the interdependencies between users, underlying [IT] infrastructure and service.

Our discussion of service availability is especially relevant given the Navy's increased interest in a service-oriented architecture (SOA) for its IT networks.

User-Perceived Service Availability

Real-world availability is user-perceived. We assert that user perception is affected by whether or not a requested service performs its function. For example, does my email work? Did my email get through? (Note that the behavior of the user himself affects perception.) For shipboard users, there are numerous services that are relevant: video teleconference (VTC), chat, defense messaging, and many others.

As Table 2.1 illustrates, network architectures are complex and can be described at different layers.

Table 2.1
Approximate Network Layers

Layers	Example Components at Each Layer
Service layer	
Application layer	
Protocol layer	IPv4, IPv6, routers/routing schemes
Physical layer	Fiber, cables, RF channels

Users perceive network availability at the service level. In contrast, the Navy tracks availability data (via NSWC-Corona) exclusively on hardware and equipment, which are part of physical layer of the network. The physical layer contributes to but does not directly reflect what the user will perceive.

Furthermore, the user-perceived availability of a service may vary from one user to the next. More importantly, the availability of a service may be more or less significant to one mission outcome than to another. Both of these factors, as we discuss next, are important to consider when evaluating user-perceived availability.

Measuring User-Perceived Service Availability

User-perceived service availability is the percentage of service invocations requested by a particular user that are successful (out of the total number of service invocations attempted by that user) during a given time interval.

Further, we could define total service availability as the mean value of all individual user-perceived service availabilities:

$$A_S = \frac{\sum_{i=1}^{n} A_i}{n}.$$

However, this definition assumes that all client usages of the service are equal. If we consider, in addition, usage factors u_i for each client i, we can define total service availability as

$$A_S = \sum_{i=1}^{n} A_i \times u_i,$$

where u_i represents the fraction of total service invocations made by client i. (We note that the sum of usage factors for all users or clients is equal to 1.)

Measuring the usage factor u_i across the fleet is not a trivial task. Fortunately, there are already efforts underway that collect precisely this sort of data. The Information Technology Readiness Review involves the installation of service monitors in target servers. These monitors produce a count of the total number of invocations made to a particular server. In the future, similar service monitors will be required in target client computers if these usage factors are to be ascertained. (These monitors come by default in the Microsoft Windows XP and Vista operating systems.) Such monitors would give the number of service invocations made by a particular client. Using these usage numbers from the client and dividing them by the total number of requests seen at the server would yield usage factors per client i, namely, u_i.

Suggested Extensions to Model User-Perceived Mission Availability

We extend the approach described above for measuring user-perceived service availability to describe a metric for measuring user-perceived mission availability. We extend the term *mission availability* to encompass not just the availability of a given service during the course of a mission, but also the impact that the service's availability has on the mission. Specifically, we define a usage factor, $u_{i,j}$, that captures the relative usage by user i of service j; the availability of service j by user i, $A_{i,j}$; the number of services that have a mission impact, M; the numbers of users, N; and a weighting factor, w_j, that describes the relative impact of service j to a given mission. The resulting user-perceived mission availability, $A_{mission}$, can then be described by

$$A_{mission} = \sum_{j=1}^{M} w_j \times \sum_{i=1}^{N} A_{i,j}\, u_{i,j}\,.$$

We note that the weighting factors w_j and $u_{i,j}$ must satisfy

$$\sum_{j=1}^{M} w_j = 1\,,$$

and

$$\sum_{i=1}^{N} u_{i,j} = 1$$

for each j.

This definition of user-perceived mission availability provides a framework that captures (1) the availability of the individual service the user is trying to invoke, (2) the relative usage factor for each service user, and (3) the relative mission impact of each service (not assessed in this study).

Measuring Software Availability

We are limited in what we can say about software availability other than it is vital, difficult to obtain, and complex. Some thoughts on how it can be calculated are as follows. Quoting EventHelix.com (2009): "Software failures can be characterized by keeping track of software defect density in the system. This number can be obtained by keeping track of historical software defect history." According to EventHelix.com, defect density[3] will depend on the following factors:

- software process used to develop the design and code (use of peer level design/ code reviews, unit testing)
- complexity of the software
- size of the software
- experience of the team developing the software
- percentage of code reused from a previous stable project
- rigor and depth of testing before product is shipped.

Other Challenges

Other challenges exist with regard to tabulating these measures for multicomponent systems. Even for strings of hardware, a straightforward calculation of availability by multiplying component values is inherently flawed. Specifically, accounting for availabilities of individual components for a system of N components by simple multiplication (e.g., $A_{oSYS} = A_{o1} \times A_{o2} \times A_{o3} \times \ldots \times A_{oN}$) assumes a number of conditions that are unlikely to hold true (Henriksen, 2006):

- The component availabilities are independent of each other.
- A single number is adequate for the entire system's availability.
- One component value will not weight the overall value to the point that the most responsible component is identifiable.
- All of the data for each component are complete and accurate.

    In particular, this approach is risky when one or more of the element availability calculations are based on incomplete or unavailable data, as in this case (Henriksen, 2006). In this report, we suggest that the difficulty in meeting all or any of the assumptions (above) can be mitigated to some extent by inserting uncertainty into the availability calculation. This is described in Chapter Six.

---

[3]   Defect density can be measured in number of defects per thousand lines of code (defects/KLOC).

Summary of Measures

The term *network dependability* is broad. It is determined, in part, by the availability and reliability of IT systems and the functions these systems provide to the user. The term *network dependability* can be used to describe (1) the ability of the Navy's IT systems to experience failures or systematic attacks without impacting users and operations, and (2) achievement of consistent behavior and predictable performance from any network access point. There is not a single universally accepted measure for dependability when it comes to IT systems and/or networks.

In the Navy, operational availability (Ao) is a metric found in many requirements documents for networks and IT systems. It is often used to assess the dependability of systems. However, a better framework for measuring network dependability should consider the following:

- the types of services available to shipboard users
- the volume of user requests for these services
- the availability of these individual services
- the impact of these services on various missions
- the relative importance of these missions (not assessed in this work).

The user-perceived service availability measure introduced in this chapter is a very useful metric for meeting these considerations for Navy networks. The user's perspective of the functionality of a service is more valuable than a measure focused only on a specific piece of hardware. In Chapter Seven, we build sufficient architectural detail into a new tool in order to provide an example of how to consider the availability of a service.

It is important to note that critical mission threads are just as dependent on the decisions and actions of human actors as they are on the mechanical operation of a given piece of equipment. In fact, the human element may be more important in this respect. Appendix B provides an example of ways to develop an understanding and measurement of the human impact on network dependability.

CHAPTER THREE

Drivers of Dependability

A number of factors drive network dependability, and hence availability and reliability. We discuss these factors in this chapter. We note the factors that are already considered by the Navy when it assesses dependability and the ones that it does not consider but should.

Summary of the Key Factors

The complexity of shipboard networks and the many factors that significantly affect the dependability of a network include

- hardware
- software applications/services
- environmental considerations
- network operations
- user (human) error
- network design and (human) process shortfalls.

It is important to note that equipment failures alone do not drive failures of IT systems. Most outages in computing systems are caused by software faults (Tokuno and Yamada, 2008; Gray and Siewiorek, 1991; Hou and Okagbaa, 2000) or human actions (Snaith, 2007; Kuhn, 1997; "Human Error Is the Primary Cause of UK Network Downtime," 2003).

In addition, the dynamics of a mission can affect network dependability, availability, and reliability. For example, geographical effects (e.g., for radio frequency transmissions, the relative position of each ship, interference, etc.) are particularly important when considering satellite communications.

PMW 750 CVN C4I CASREP Study

Within PEO C4I, a Casualty Report (CASREP) study was recently completed by PMW 750[1] using 169 CASREPs recorded aboard carriers (Conwell and Kolackovsky, 2009). The stated purpose of this study was to identify root causes and any systemic issues for most systems on the CVNs. The data were collected over a one-year period (July 2007 to July 2008).

The resulting analysis of these CASREPs indicated that most of the problems fell into nine categories, which can be considered first-order root causes. Specifically, Conwell and Kolackovsky binned these root causes as follows:

1. Hardware (37 percent, or 63 out of 169)
2. Training (16 percent, or 27 out of 169)
3. Integrated Logistics Support[2] (ILS) (12 percent, or 20 out of 169)[3]
4. Design (11 percent, or 19 out of 169)
5. Configuration Management (9 percent, or 15 out of 169)
6. Settings (4 percent, or 7 out of 169)
7. Software (4 percent, or 7 out of 169)
8. System Operational Verification Testing (SOVT) (3 percent, or 5 out of 169)
9. Other (4 percent, or 7 out of 169).

This is shown in the pie chart in Figure 3.1. What is clear from this analysis is that nearly two-thirds of the CASREPs do not directly involve equipment failures. This fact is in spite of a data skew: Many of the systems surveyed have no software component.

If we create a meta-category called "human and human-defined processes" and include training, design, SOVT, and ILS within it, then that meta-category accounts for 42 percent of CASREPs using the study data. If we include software, configuration management, and settings into a more general "software" category, we find that 16 percent of CASREPs fall into this bin. Again, not all systems observed in this study had a software component.

Conwell and Kolackovsky's observations are consistent with other literature that finds more than 50 percent of failures in networked IT environments to be caused by human error (Snaith, 2007; Kuhn, 1997; "Human Error Is the Primary Cause of UK Network Downtime," 2003). Other studies flag software: Hou and Okogbaa (2000) gathered industry data points to conclude that software is a major cause of failures in networks rather than just hardware for systems that involve both.

---

[1]   PMW 750 (Carrier and Air C4I Integration Program Office) is an organization with PEO C4I that serves as the integrator for C4I systems.

[2]   This represents issues related to part supplies, such as the wrong part being sent or not having enough spares.

[3]   The ILS category includes parts not being onboard and technical manuals not being up to date or available.

Figure 3.1
Root Causes of CASREPs

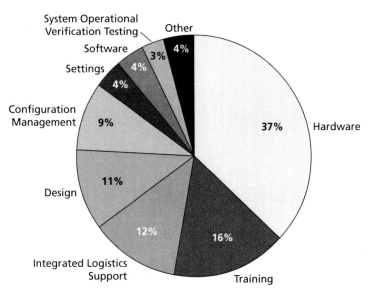

SOURCE: Conwell and Kolackovsky, 2009.
RAND *MG1003-3.1*

A review of outage time from Conwell and Kolackovsky (2009) is shown in Table 3.1. From this perspective, the broad category of human and human-defined processes still makes up most of the outage time documented in the study.

The main takeaway from Figure 3.1 and Table 3.1 is that the majority of root causes are not tied to a hardware failure. Certainly, additional bins can be identified that span several of the bins used by Conwell and Kolackovsky. Reportedly, the PMW 160 Fleet Support Team IPT considers that "software configuration" issues are a main root cause. This would be a new bin (not listed above) that could span the existing training, setting, configuration management, and software root cause bins identified in the PMW 750 CVN C4I CASREP study.

Hardware Failure

NSWC-Corona captures equipment uptime and downtime of key hardware. In general, hardware component availability rates are easier to obtain than other types (e.g., software availability rates). As the CASREP CVN C4I study points out, hardware equipment failures are considered a major culprit in network IT failures. Indeed, a traditional calculation of a system Ao is to simply multiply the availabilities of its dependent components.

Table 3.1
Outage Time

Cause	Outage Time (%)	Outage Time (days)
Hardware	32	1,775
Training	20	1,126
Design	15	829
ILS	11	612
Configuration Management	8	446
Settings	3	161
SOVT	4	243
Software	2	93
Other	4	215

SOURCE: Based on Conwell and Kolackovsky, 2009.

NOTE: Not all systems observed in this study had a significant software component. Examples of systems tracked that did not include significant software involve satellite and radio equipment (including antennas, terminals, WSC-3, and WSC-8).

However, hardware issues are not the only significant factor. While it is convenient to focus more on hardware issues (since more data are available), hardware-only modeling is insufficient for modeling network dependability. In addition to hardware, two big drivers of availability are (1) flawed human processes and human error and (2) software problems.

Software Problems and Human Error

Industry data point to software failure as being a major cause of failures in networks and computing systems rather than hardware failure (Hou and Okogbaa, 2000; Tokuno and Yamada, 2008). And ITRR data from CVNs show that a large proportion of CASREPs were caused by human error, not equipment failure.

Some reports suggest that more than 50 percent of failures in networked IT environments are caused by human error (Snaith, 2007; "Human Error Is the Primary Cause of UK Network Downtime," 2003). This finding does not differ from other IT innovations from decades past, as shown in Table 3.2.

It is important to note that many of the academic publications on the topic of human error in network reliability analysis (e.g., Reeder and Maxion, 2005; Maxion

**Table 3.2
Sources of Failure in the Public Switched
Telephone Network**

Source of Network Failure	Percentage
Human error: company	25
Human error: others	24
Hardware failures	19
Software failures	14
Acts of nature	11
Overloads	6
Vandalism	1

SOURCE: Based on data in Kuhn (1997).

and Reeder, 2005) focus on the issue of interface design.[4] In our assessment, so should the Navy.

Maintainers

There are specific human roles that result in failures in networked IT environments. Users and maintainers are two roles in particular that have failure studies devoted to them. Human errors committed by network equipment maintainers cause the largest portion of all network outages and downtime. Specifically, the industry average for the percentage of downtime caused by procedural errors is 50 percent (O'Brien, 2007).

Mission Dynamics

Finally, future analysis needs to consider the dynamics of a mission when considering network dependability. This includes accounting for geographical effects (e.g., line of sight for radio frequency transmissions, the relative position of each ship, interference,[5] etc.) which are nontrivial. These effects are particularly important when considering satellite communications.

---

[4]   Quoting Reeder and Maxion (2005):

Delays and errors are the frequent consequences of people having difficulty with a user interface. Such delays and errors can result in severe problems, particularly for mission-critical applications in which speed and accuracy are of the essence. User difficulty is often caused by interface-design defects that confuse or mislead users.

[5]   Consider the case where super-high-frequency (SHF) communication is degraded: The effect of interference from neighboring ships becomes a significant factor affecting network availability. Some ships have been known to block out a section of their emitters to mitigate this interference.

Conclusions

To summarize, in addition to hardware-equipment failures, two significant drivers of dependability that cannot be ignored are (1) flawed human processes and human error and (2) software problems. Estimates on the magnitude of their impact, relative to hardware equipment failures, vary, but many studies cited in this chapter suggest that it is appreciable. Finally, the role of interface design between the user and the IT device is also relevant.

Understanding the specifics of dependability and availability is highly complex. Problem formulations can become intractable. A pragmatic approach is required.

CHAPTER FOUR

Data Sources

There are a number of data sources that track metrics (e.g., Ao) that can provide valuable insight into network dependability. Among these data sources are:

1. **NSWC-Corona database.** NSWC-Corona is the division of NSWC that provides historical data for MTBF, MTTR, and MLDT for components relevant to the ASW mission.
2. **CASREPs.** Casualty reports are failure reports that someone fills out on a per-case basis. They can involve the failure of a hardware component. The CASREPs go out to ISEAs, and they in turn send out a replacement part or an engineer to make a repair.
3. **Fleet Systems Engineering Team (FSET) Reports.** Microsoft Word format.
4. **ITRR**. Microsoft Word format. This is an actual two-to-four-week evaluation of a ship's IT systems and personnel. It may involve two to four evaluators onboard while the ship is underway. Written tests of the IT staff as well as automated tests of the systems are done.
5. **Board of Inspection and Survey (INSURV).** Text format.
6. **Deploying Group Systems Integration Testing (DGSIT).** Text format.
7. **Requirements documents.** In this report, exemplar analysis presented is based largely on data from NSWC-Corona and some requirements documents (when Corona data are not available).

Corona Database

Examples of components we gathered data for from NSWC-Corona are listed in Tables 4.1–4.3.

Much of these data were based on all recorded failures of systems and components on a per-hull basis during a four-year time period. Hulls considered were CVNs and DDGs (as these were modeled in the ASW mission). We used these data in our tool in place of the requirements data previously used in the original SPAWAR 5.1.1 model. Unfortunately, historical data were not available for all ASW model elements

Table 4.1
NSWC-Corona Availability Data Used

Component	Comments on Data Collected	Data Source
DMR	FYs 2005, 2006, 2007, and Q2 2008	NSWC-Corona
ADNS HW	2007 estimate	NSWC-Corona
ADNS SW	2007 estimate	NSWC-Corona
ISNS	FY2008, June, July, August, September	NSWC-Corona
WSC-6	FYs 2005, 2006, 2007, and Q3 2008	NSWC-Corona
USC-38	FYs 2005, 2006, 2007, and Q3 2008	NSWC-Corona
WSC-8	FYs 2005, 2006, 2007, and Q3 2008	NSWC-Corona
CDLMS	FYs 2005, 2006, 2007, and Q3 2008	NSWC-Corona
HFRG	FYs 2005, 2006, 2007, and Q3 2008	NSWC-Corona
in place of Link 11 DTS)	FYs 2006, 2007, and 2008	NSWC-Corona
KIV-7	Have only 1 data point, year = ?	STF
TELEPORT TRANSEC INTERNAL	Have only 1 data point, year = ?	STF
TELEPORT IAD	Have only 1 data point, year = ?	STF
Automated Digital Multiplexing System (ADMS)	Have 2 data points, year = ?	Using AN/FCC-100 Multiplexer and Nextera Multiplexer ST-1000 data from STF as 2 data points for ADMS

(systems and components), and hence, for these elements, we continued to use requirements data. This is a partial list.

For all elements in the ASW mission, regardless of whether historical data were available or not, we found it advisable to create probability distributions for MTBFs and mean down times (MDTs).

In addition, we learned that Regional Maintenance Centers (RMCs) and port engineers may have their own trouble ticket systems or databases that would further add to the current stock of availability data.

Platform Variability

Results can be different from one platform to the next. However, preliminary examination of historical Determination of Availability (DOA) data[1] shows that this is not the case. Interviews with PMW 160 personnel reveal that cost and human error are two major reasons for the different performance of ADNS systems across the same plat-

---

[1]   There are individual efforts within PEO C4I to measure operational availability. PMW 160's DOA effort is one example. The objective of the DOA effort is to enable a ship's operational availability of ADNS to be measured and reported, where operational availability refers to three key parameters: (1) ADNS up time, (2) bandwidth availability, and (3) traffic transported.

Table 4.2
Data Sources for Systems and Components in
ASW Mission

Data Source	System or Component
Historical data obtained from NSWC-Corona	DMR on DDGs
	DMR on CVNs
	ADNS HW
	ADNS SW
	ISNS
	CDLMS on CVNs
	CDLMS on DDGs
	HFRG on CVNs
	OE-82
	ADMS on CVNs
	ADMS on DDGs
	LINK 16
	WSC-6 on CVNs
	WSC-6 on DDGs
	USC-38 on CVNs
	USC-38 on DDGs
	WSC-8 on CVNs
Requirements data (in absence of historical data)	UFO
	TVS
	COP
	DSCS
	MILSTAR
	DATMS
	HSGR
	FLT NOC TSW +
	CND
	DISN
	MOC
	EARTH STATION
	INTELSAT
	MISSION COMPUTER
	LINK DTS
	HF
	MIDS
	COMBAT SYSTEM
	KIV-7 COMSEC
	KIV-7 TRANSEC
	TELEPORT TRANSEC INTERNAL
	TELEPORT IAD

Table 4.3
Data Sources for Systems and Components in Chat and COP Services

Data Source	System or Component
Historical data obtained from NSWC-Corona	KG175 TACLANE
	KG-84
	EBEM-MD-1324 Satellite Modem (used to model MD-1366 Modem)
	WSC-6 Terminal
	WSC-6
	WSC-8 Terminal
	WSC-8
Manufacturer's specifications used (in absence of historical data)	Alcatel 4024 Access Layer Switch
	Alcatel OnniSwitch 9WX Access Layer Switch
	Alcatel Omni Switch/Router OS/R
	AN/FCC-100 Multiplexer
	Cisco 2611XM Router
	Cisco 3745 Router 3750
	Cisco 6506 Switch 6500
	Cisco 4500 Router
	Cisco 3845 Router
	Cisco 3560 Switch
	Cisco 2950 Switch
	Cisco 2960 Switch
	Nextera Multiplexer ST-1000
	AN/USQ-155(V) TVS
	KIV-7M
	KIV-84A
	KIV-7SHB
	TRANSEC KG-194
	KIV-7
	EBEM-MD-1030B Satellite Modem
	MIDAS
	EHF Follow-On-Terminal (FOT)
	Patch Panel EIA-530
	Sidewinder G2 1100 D Firewalls
	HP DL385
	Packeteer 1700
	Packeteer 10000
	DSCS
	MILSTAR
	INTELSAT
	INMARSAT
	UFO

forms. Sometimes, commanders choose to shut down the ADNS system for a variety of potential reasons (e.g., drills, maintenance). Other times, ADNS operators incorrectly configure the system and cause an outage. In the former case, ADNS analysts should review ships' logs to account for manual shutdown of the system so that these time periods are not included in the availability analysis.

Some data were sourced from requirements documents and manufacturers specifications when no fleet data was available. The extent to which this was done is shown in Tables 4.2 and 4.3. When required, distributions were still employed to account for uncertainty.

Observations and Recommendations

Fuse the Data

Different organizations pay attention to different types of data, for better or for worse. Consider the results of a 2007 Gartner survey regarding how availability data are tracked, which indicate that many in industry use both data regarding the IT infrastructure (similar to data Corona provides) and RTTS data. As shown in the Table 4.4, the top strategy is to use a "combination of service desk and IT infrastructure measurements" (28 percent).

The number of data sources presents challenges (and opportunities). Key questions that must be addressed are

- How does an analyst consolidate all these disparate data sources?
- How do we address the varying formats? For example, some data sources, such as FSET, ITRR, INSURV, and DGSIT, are in Microsoft Word or text format, making them difficult to analyze.

Table 4.4
Findings of Industry Survey on How Availability Data Are Tracked

Response	Percentage
Combination of service desk and IT infrastructure measurements	28
Service desk metrics (user calls, outage records, trouble-ticket MTTR)	22
End-to-end application transaction response time (top-down)	16
Aggregating IT infrastructure component availability (bottom-up)	16
IT infrastructure component availability only (not aggregated together)	11
We don't need availability metrics	7

SOURCE: Based on data in Gartner (2007).

The way ahead probably requires the creation of a new database to be populated manually from these reports to make them usable. Finally, some data sources are more closely guarded than others, and thus the cultural barriers must be deconstructed.

Create and Utilize Standards

The multiple data sources can provide insight if fused, but, due to their varying formats, they did not easily lend themselves to a standardized data-extraction process. We believe the Navy would be well served to create a standard for reporting any form of availability or reliability-related issues. Although text documents are informative, they are not easily analyzed in bulk, and an analyst would have to create a brand new database and populate it in order to investigate these results. The XML format is an example of a standardized data format that is gaining wide acceptance in industry. If the various text-based reports (ITRR, DGSITs, INSURVS, etc.) were to make their data available in XML format, analysts could more easily import such data into a database rather than manually entering them by hand. Alternatively, authors of these reports could directly enter data into the Remedy database. This latter option would be the preferable approach, as the Navy already has a large adoption of the Remedy application. Also, Remedy has built-in analysis tools, eliminating the need for the creation or purchase of third-party tools to analyze the data.

CHAPTER FIVE

Modeling a Service: Examples Using SameTime Chat and the COP Service

We describe a modeling effort to account for the availability of individual components that support the delivery of a service. At the component level, this is a black box approach. This approach is an example of how additional resolution can be applied without attempting to model internal details (e.g., protocols supporting). (Concepts such as quality of service, prioritization, and bandwidth are not treated.)

Six-Step Process

The creation of a service model involves a six-step process similar to that developed by Malek et al. (2008):[1]

1. Identify the important services and define the required availability of each service.
2. Collect infrastructure data responsible for the delivery of the service. This includes equipment and other dependent services on the network along with their availability statistics.
3. Transform the infrastructure/network diagram into a connectivity graph. This integrates computing and communication infrastructure by adding communication links to the computing nodes.
4. Map the steps of the service execution to the connectivity graph. Each step in the execution has its source S, which initiates the step, and destination D, which performs the step, in the connectivity graph. The purpose of this step is to find all possible paths between S and D. We transform the individual source-destination paths into Boolean equations by applying & (AND) operators between nodes that belong to the same path, and the || (OR) operator between

---

[1] The RAND approach deviates slightly from the Malek et al. seven-step approach in that Malek et al. intended to generalize their methodology as much as possible. RAND adapted Malek et al.'s methodology to Naval networks and tailored the seven steps to be more applicable to the task at hand.

parallel paths. For instance, consider an email client attempting to contact an email server, as depicted in Figure 5.1. The client can access the server using the path that traverses routers R1 and R2, or alternatively, it can select the path that traverses only router R3. We express this as:

$$CLIENT\&(R1\&R2)\&SERVER\|CLIENT\&R3\&SERVER =$$
$$CLIENT\&((R1\&R2)\|R3)\&SERVER$$

5. Transform the mapped service description from step 4 into a formal reliability block diagram.
6. Insert the resulting reliability block diagram into the RAND Network Availability Modeling Tool and compute the availability of the system.

Step 4 in this process requires the identification of all paths between a given source and destination pair. The complexity associated with this step depends on the network infrastructure—as the number of loops in the topology increases, the algorithm complexity grows. Hypothetically, this complexity is prohibitive, as Malek et al. point out, because "in the worst case, when a graph is a complete graph (each two vertices are connected by an edge), the time/space complexity of the algorithm is $O(n!)$" (2008, p. 7).

This complexity is not a major concern in IT networks, because most networks in this realm are computationally feasible. Specifically, most networks in the IT realm consist of tree-like structures with limited numbers of loops in the topology. These loops consist mainly of routers, and their count is generally limited. A moderate number of network switches connect to each router. It is the host systems connected to these switches that create the tree-like structures. Such structures yield a mathematically approachable problem with limited complexity.

Figure 5.1
Multiple Paths Between Clients and Servers

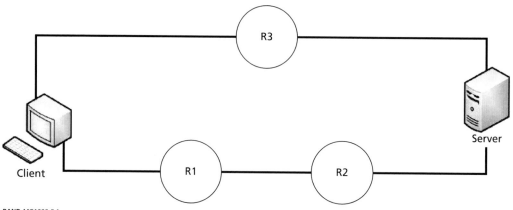

Case Studies

Case Study: Modeling the SameTime Chat Service in CENTRIXS

We selected SameTime chat in the Combined Enterprise Regional Information System (CENTRIXS) in a CVN platform as our initial case study due to its standardized implementation across the fleet and its importance to both U.S. and coalition forces.

Steps 2 and 3 in the six-step approach described above—collecting the necessary data/network diagrams from each respective department to craft a complete and accurate service model—proved to be a major challenge for this effort. The primary source of difficulty was the absence of a comprehensive source of network architecture across the Navy. Each PMW or technical department, such as PMW 160 or PMW 790, focused on the architecture of their respective enclave (rightfully so), and although it was straightforward to obtain equipment inventories and network diagrams for individual units, integrating them into a complete picture that describes the SameTime chat service in its entirety demanded a great level of time and coordination among the numerous stakeholders. Later, in the recommendations section of this report, we strongly suggest that the Navy create a central repository of network diagrams to facilitate future reliability analyses. The Navy's creation of a website (not available to the general public) intended specifically for ISEAs is a positive step in this direction, but, at the time of this writing, it still lacks shore diagrams, which would enhance the usefulness of the site. There is a PEO C4I wide effort called SAILOR that this will be rolled into.

The completed network diagram for the SameTime chat service is depicted in Figures 5.2–5.4. For these figures, a portion of SameTime chat was taken from CENTRIXS diagrams, and a portion was taken from ADNS diagrams. Given the architecture depicted in the figure, we created a dependency chain based on the network diagrams. Due to the large size of the SameTime chat service, we divided the different process executions as follows, using the following abbreviations:

2811 = Cisco 2811 Router
3560 = Cisco 3560 TACLANE Switch
3745 = Cisco 3745 Router
3845 = Cisco 3845 Router
6506 = Cisco 6506 ADNS Policy Switch
ALC = Alcatel 4024 CENTRIXS switch
EBEM = EBEM MD-1366 Satellite Modem
FW = CENTRIXS Sidewinder G2 Firewall
ISSR = Cisco 3845 CENTRIXS Router
MUX = Multiplexer (Shore)
PACK = Packeteer 1000
PS = Packetshaper
P/P = Patch Panel
SHOR = Shore Satellites.

Figure 5.2
Network Diagram for SameTime Chat (1 of 3)

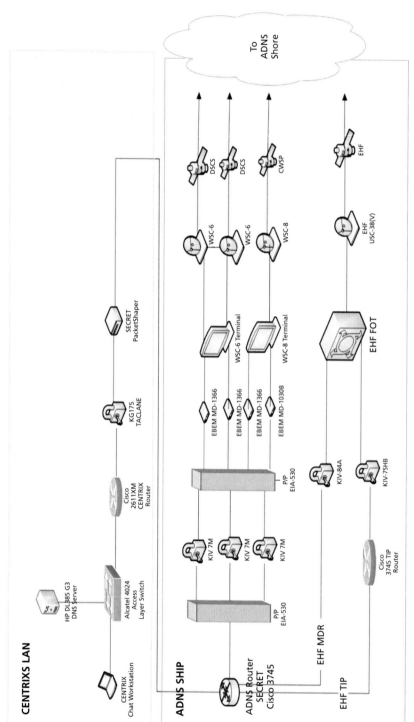

RAND *MG1003-5.2*

Figure 5.3
Network Diagram for SameTime Chat (2 of 3)

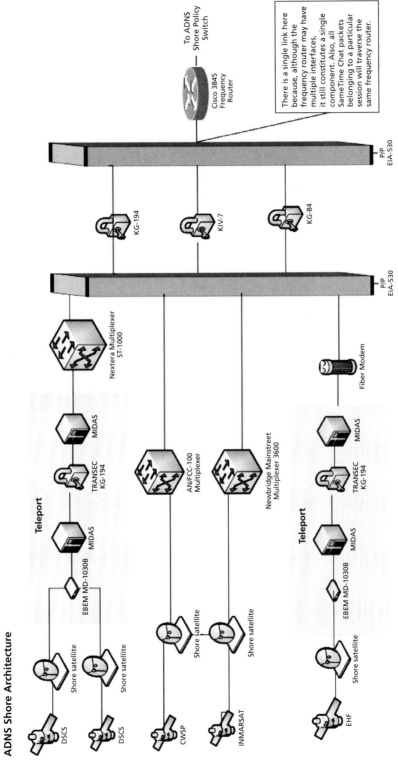

Figure 5.4
Network Diagram for SameTime Chat (3 of 3)

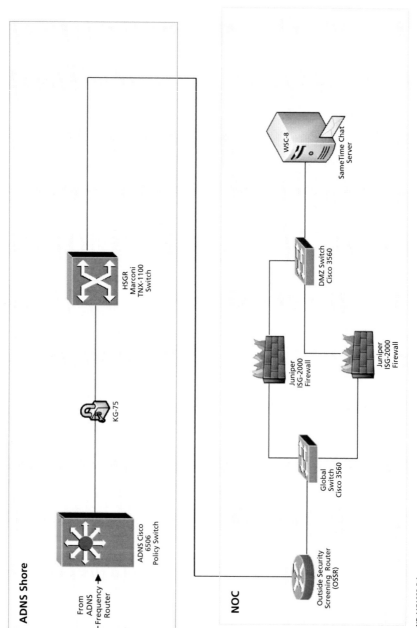

RAND MG1003-5.4

In order to communicate with the SameTime chat server located in the NOC, the chat workstation has to resolve the address of the server with the local DNS server:

CL -> DNS:
    CL&ALC&DNS

Once this process is complete, the workstation can proceed to contact the chat server, first through the ADNS ship infrastructure:

CL -> ADNS:
    CL&ALC&2811&KG175&PS1&3745

The ADNS infrastructure branches into several parallel paths, and to simplify the final expression, we consider each individual path separately:

ADNS1 -> SATCOM:
    3745&P/P1&(KIV1‖KIV2‖KIV3)&P/P2

The SATCOM paths then divide into three possible paths:

SATCOM1 -> SHORE:
    P/P2&(EBEM1‖EBEM2)&WSC6T&((WSC61&DSCS1&SHOR1)‖
    (WSC62&DSCS2& SHOR2))&EBEM-SHOR1&MIDAS1&KG194&
    MUX&P/P3

SATCOM2 -> SHORE:
    P/P2&(EBEM3‖EBEM4)&WSC8T&WSC8V2&
    ((CWSP&SHOR3&FCC1000)‖
    (INMARSAT&SHOR4&NEWBRIDGE))&P/P3

ADNS2 -> EHF:
    3745&(KIV8‖(3745#2&KIV7SHB))&EHF-FOT&USC38(V)&EHF

EHF -> SHORE:
    EHF&SHOR5&EBEM-SHOR2&MIDAS3&KG194&MIDAS4&P/P3

Finally, the last serial chain describes the ADNS Shore infrastructure delivering chat traffic to the CENTRIXS SameTime chat server:

SHORE -> CHAT:
    P/P3&(KG194‖KIV7‖KG84)&P/P4&3845&6506&PACK&PS2&3845&
    3560&KG175&ISSR&PS3&3560&(FW1‖FW2)&3560&CHAT

For the successful execution of the SameTime chat service, all these steps must be performed. The resulting expression is simplified by applying the idempotence, associativity, and distributivity rules of the & and || operators:

CL -> CHAT:
(CL -> DNS)&(CL -> ADNS)&(((ADNS1 -> SATCOM)&((SATCOM1
-> SHORE)||(SATCOM2 -> SHORE))||((ADNS2 -> EHF)&(EHF ->
SHORE)))&(SHORE -> CHAT)

We transform this expression into an RBD, as shown in Figures 5.5–5.7. (Software portions are not modeled in this example but doing so is recommended for future work, as discussed later in this report.) We then inserted this RBD into the Network Availability Modeling Tool for analysis. Results of running the tool on the SameTime chat service are discussed in a later chapter.

Case Study: Modeling the COP Service in the Third Fleet

Upon completion of the SameTime chat model, PEO C4I requested that RAND study COP as its next case study. Due to the large breadth and scope of the COP service, we isolated the study to Blue Force Tracking within GCCS.

The COP service is inherently different from the SameTime chat service and other conventional client/service architectures due to its dependence on the synchronization of databases spread throughout the fleet. In the chat service, as in email and web access services, a client initiates a request, which is then forwarded to the server. In the COP service, a client may also initiate requests within the GCCS application. The integrity of the program, however, depends heavily on its ability to get updated information from different COP servers. The COP Synchronization Tool (CST) is the preferred method of exchanging data between COP servers:

> A critical component of COP operations is the initialization, maintenance, and management of the CST network supporting real-time GCCS track database exchange throughout the COP federation. CST enables each node to receive raw and processed track information and distribute the results of track correlation and fusion throughout the CST network. (Chairman of the Joint Chiefs of Staff, 2008)

Whereas most client/server architectures are "pull" designs—that is, a client "pulls" a resource from a server—COP is a combination of both a "pull" and "push" architecture—the "push" being when one database is updated and sends the updated data to all other COP servers.

With the guidance of PMW 150 and COMPACFLT engineers, RAND gathered the necessary information to model the COP service from the USS *Nimitz* in the Third Fleet. Due to the unavailability of data from the COP server in Camp Smith in

Figure 5.5
Reliability Block Diagram for SameTime Chat Service (1 of 3)

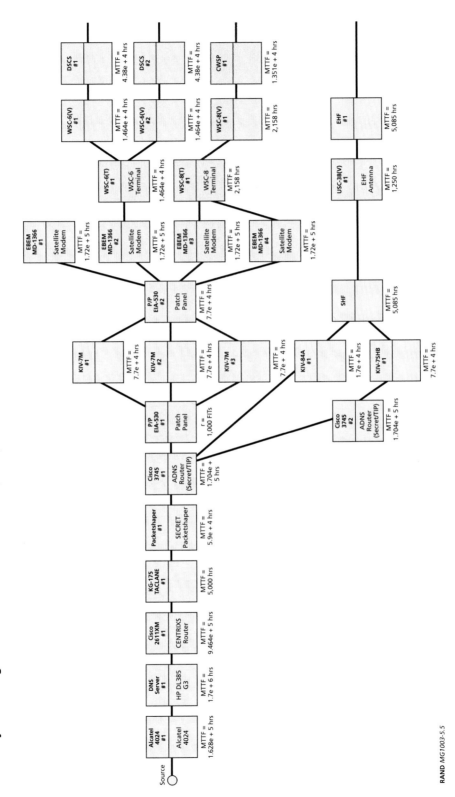

Figure 5.6
Reliability Block Diagram for SameTime Chat Service (2 of 3)

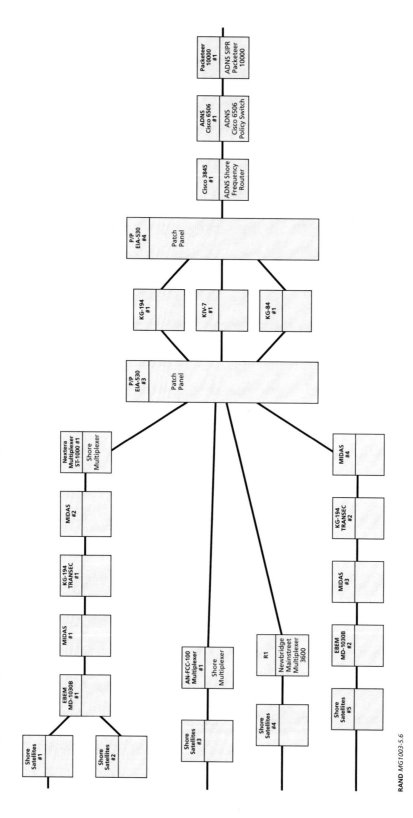

RAND *MG1003-5.6*

Figure 5.7
Reliability Block Diagram for SameTime Chat Service (3 of 3)

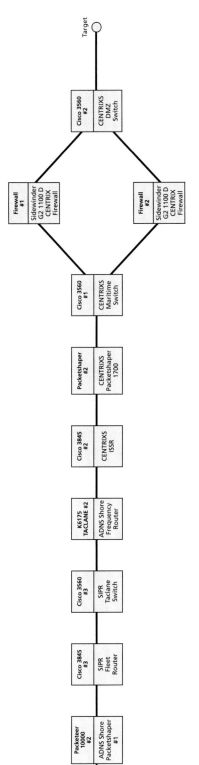

RAND *MG1003-5.7*

Hawaii, this study addressed a subset of the complete COP deployment from the USS *Nimitz*. Figure 5.8 shows the area of study.

Using the same approach described for SameTime chat, we created a comprehensive network diagram for the COP service, reduced it to a connectivity expression, and then translated it into an RBD. The ADNS backbone for both the SameTime chat and the COP services were identical, as they were both taken for CVN platforms. The major differences in the network diagrams were in the shipboard ISNS architecture and the shore connectivity. Appendix C contains the complete network diagram for COP. Later in the report, we present the results of the simulation.

Observations and Conclusions

In a previous chapter, we suggested that whole services (from provider to user) should be modeled. The advantage of this approach is that a more accurate representation of what users experience is represented. Furthermore, this approach is consistent with the CANES paradigm of service-oriented architecture (SOA).

Figure 5.8
RAND COP Model Includes Subset of Total Architecture, as Indicated in the Area Circled in Red

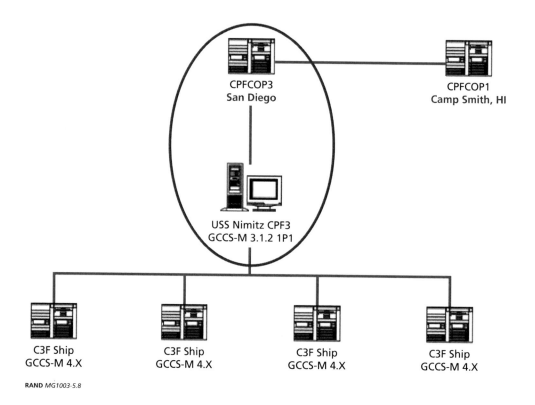

RAND *MG1003-5.8*

We chose several mission-critical services (e.g., chat, secure voice) as an example analysis. A key challenge for this effort was collecting the necessary data/network diagrams from each respective department to craft a complete and accurate service model. For example, we found that representing the chat service required a fusion of separate architectures, as illustrated in Figure 5.9.

Figure 5.9 shows that services cross program boundaries. Thus, in order to come up with a complete model that fully describes a particular service, a considerable effort is required. For example, there is a need to acquire multiple (separate) architecture diagrams. In the case of secure chat, we have to understand which different departments are responsible for the delivery of the chat service. In this case, the relevant information was contained in CENTRIXS, ADNS, and NOC engineering data. In summary, there is a need to consolidate data from multiple sources/departments to come up with a more complete service architecture.

Figure 5.9
Challenge—Services Cross Program Boundaries

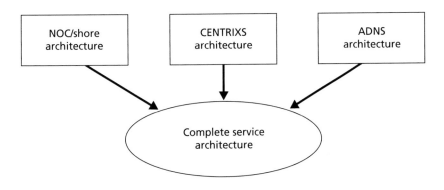

RAND *MG1003-5.9*

CHAPTER SIX

A New and Improved Tool to Calculate Ao

Overview of RAND's Network Availability Modeling Tool

As discussed earlier in this report, the PEO C4I requested that RAND evaluate the SPAWAR 5.1.1 spreadsheet model, which evaluated availability for the ASW mission and is described in Appendix A, and suggest improvements. In response to this request, RAND built upon the existing spreadsheet model to develop a network availability modeling tool with greatly enhanced capabilities from that of the original SPAWAR 5.1.1 spreadsheet. The front end of the tool is shown in Figure 6.1. Multiple menus also exist, but they are not shown in this figure. As suggested by the items in the main menu, this Excel-based tool can accommodate models of equipment strings for

Figure 6.1
Main Menu of the Tool

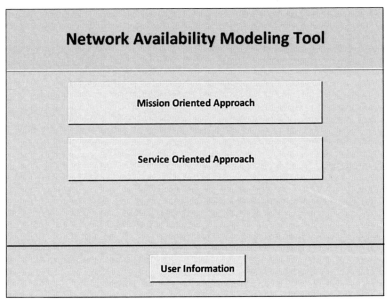

Network Availability Modeling Tool

Mission Oriented Approach

Service Oriented Approach

User Information

RAND *MG1003-6.1*

the ASW mission (a mission-oriented approach to analysis). In addition, the tool can analyze two specific services (chat and COP) as part of an approach to model services.

The primary enhancements (relative to the initial SPAWAR tool) are as follows:

1. We added stochastic models of all systems' and components' performance metrics, namely MTBF and MDT, as opposed to the hard-coded requirements data that were previously used.
2. When available, we fit historical data from NSWC-Corona to probability distributions to define the stochastic model for system and component MTBF and MDT. For those systems and components without historical data, we fit requirements data to a probability distribution with user-modifiable parameters.
3. We expanded the original spreadsheet model with a Monte Carlo add-on package to Excel (@RISK) and developed a user friendly front-end GUI that enables the user to navigate the tool without specific knowledge of Excel or @RISK.
4. We added the capability for the user to visualize both the availability histogram that results from the Monte Carlo simulations as well as a sensitivity analysis plot that informs the user about which components or systems have the largest impact in terms of their effect on the resulting mission availability. Both the availability histogram and the sensitivity analysis plot can be visualized for the entire mission or on a per–equipment string basis.
5. We added models for both SameTime chat and COP services to better evaluate user-perceived service availability. These models are discussed further in Chapter Seven, with subsequent analysis of our results in Chapter Eight.

Stochastic Models Used for MTBF and MDT

We obtained historical data for MTBFs and MDTs for many of the systems and components used in the ASW mission from NSWC-Corona. In particular, we obtained data for DMR, WSC-6, USC-38, WSC-8, CDLMS, HFRG, ADMS, OE-82, ADNS, and ISNS. Much of these data were based on all recorded failures of systems and components on a per-hull basis during a four-year time period. Hulls considered were CVNs and DDGs, as these were modeled in the ASW mission. We used these data in our tool in place of the requirements data previously used in the original SPAWAR 5.1.1 model. Unfortunately, historical data were not available for all ASW model elements (systems and components), and hence, for these elements, we continued to use requirements data. The list of elements for which we obtained historical data and for which we used requirements data is summarized in Chapter Four. For all elements in the ASW mission, regardless of whether historical data were available or not, their MTBFs and MDTs were fit to probability distributions. We describe these distributions, as well as the options available to the user when selecting them, in more detail below.

Operating RAND's Tool in "Component" Mode

The RAND tool has two primary modes, component and equipment string. In component mode, the user can modify parameters for an individual element (component or system). The user selects both the equipment string and the individual element in the selected equipment string from a pull-down menu. As seen in Figure 6.2, the user has selected DMR on DDGs in the "voice" equipment string. We describe below in further detail those parameters of the selected element that the user has control of.

User Can Modify PERT's Parameters

For those elements with limited or no historical data, the user fits the data to a PERT distribution, manually controlling its parameters. PERT distributions (or Beta PERT distributions) are frequently used to model experts' opinions and have many desirable properties. For instance, they are specified by three intuitive parameters: minimum, most likely, and maximum values. Their mean is four times more sensitive to the most likely value than to the minimum or maximum values. The standard deviation is also less sensitive to estimates of extreme values than it is for a triangular distribution, and the PERT distribution does not overestimate the mean value when the maximum value is large. Figure 6.3 shows a PERT distribution with a minimum value of 0, a most likely value of 0.2, and a maximum value of 1.

In the RAND tool, the user can vary all three of PERT's parameters—the minimum, most likely, and maximum values—for both MTBF and MDT for all components and systems in the model by adjusting scroll bars on the front end of the tool. Figure 6.2 shows a partial view of this front end, focusing on the part that gives the user control over PERT's parameters for both MTBF and MDT for the selected component, which in this example is DMR on DDGs in the "voice" equipment string. The tool shows the minimum, most likely, and maximum values from the historical data (or requirements data if historical data are not available) in the "Historical Data" box. Below it, in the "User Modification" section, the user may adjust scroll bars to control the parameters of the PERT distribution that these data will be fit to. In this example, we see that the user has selected to both decrease the minimum and increase the maximum MTBF and MDT values by 15 percent. After making their modifications, the user clicks the "update" button to see the resulting parameters in the "Resulting User Modified Values" box.

User Can Manually Select a Distribution or Opt to Use a "Best-Fit" Distribution

For those elements in the model with sufficient historical data, the user can choose between a PERT distribution, a manually selected distribution, and a "best-fit" distribution. To manually select a distribution, the user highlights the "select distribution" radio button and then selects from a pull-down menu their distribution of choice. If the selected distribution does not give a valid fit to the data, the best-fit distribution is used instead. Figure 6.4 shows a partial front-end view of RAND's tool, highlighting

Figure 6.2
RAND Tool Front End Component Mode: Selecting a PERT Distribution

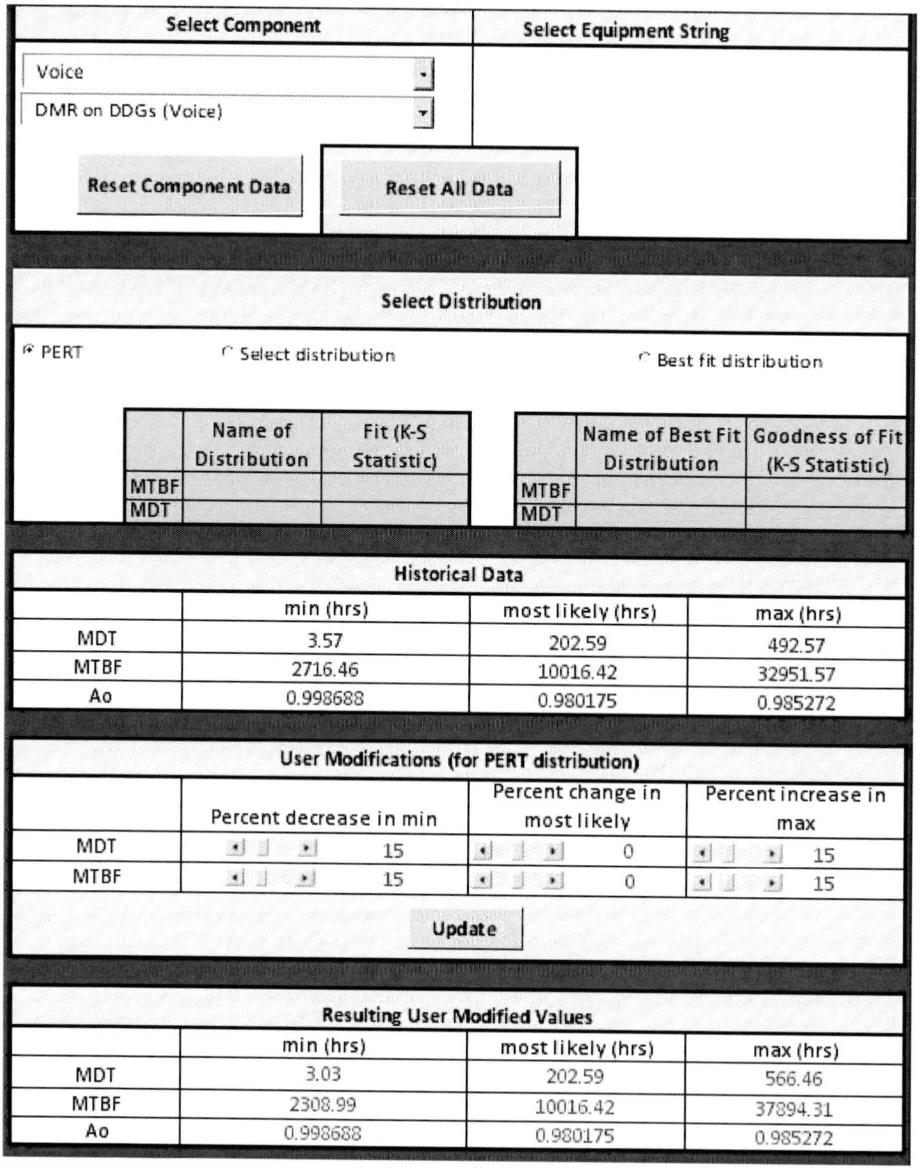

Select Component	Select Equipment String
Voice	
DMR on DDGs (Voice)	
Reset Component Data **Reset All Data**	

Select Distribution

⊙ PERT   ○ Select distribution   ○ Best fit distribution

	Name of Distribution	Fit (K-S Statistic)		Name of Best Fit Distribution	Goodness of Fit (K-S Statistic)
MTBF			MTBF		
MDT			MDT		

Historical Data

	min (hrs)	most likely (hrs)	max (hrs)
MDT	3.57	202.59	492.57
MTBF	2716.46	10016.42	32951.57
Ao	0.998688	0.980175	0.985272

User Modifications (for PERT distribution)

	Percent decrease in min	Percent change in most likely	Percent increase in max
MDT	15	0	15
MTBF	15	0	15

Update

Resulting User Modified Values

	min (hrs)	most likely (hrs)	max (hrs)
MDT	3.03	202.59	566.46
MTBF	2308.99	10016.42	37894.31
Ao	0.998688	0.980175	0.985272

RAND *MG1003-6.2*

the case in which the user has opted to manually select a distribution, in this case a beta distribution. Similarly, the user may opt to use a "best-fit" distribution by selecting the "best fit distribution" radio button, as seen in Figure 6.5. The best-fit option enables the use of that distribution for both MTBF and MDT that "best explains" the data according to some metric. We refer to this metric as the *goodness-of-fit* metric and describe it in more detail below.

Figure 6.3
PERT Distribution with Parameters 0, 0.2, and 1

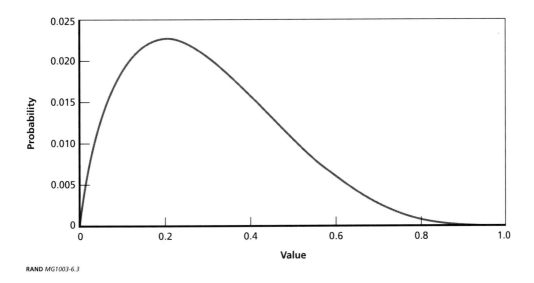

RAND *MG1003-6.3*

Figure 6.4
Front End of RAND's Tool in Component Mode: Manually Selecting a Distribution

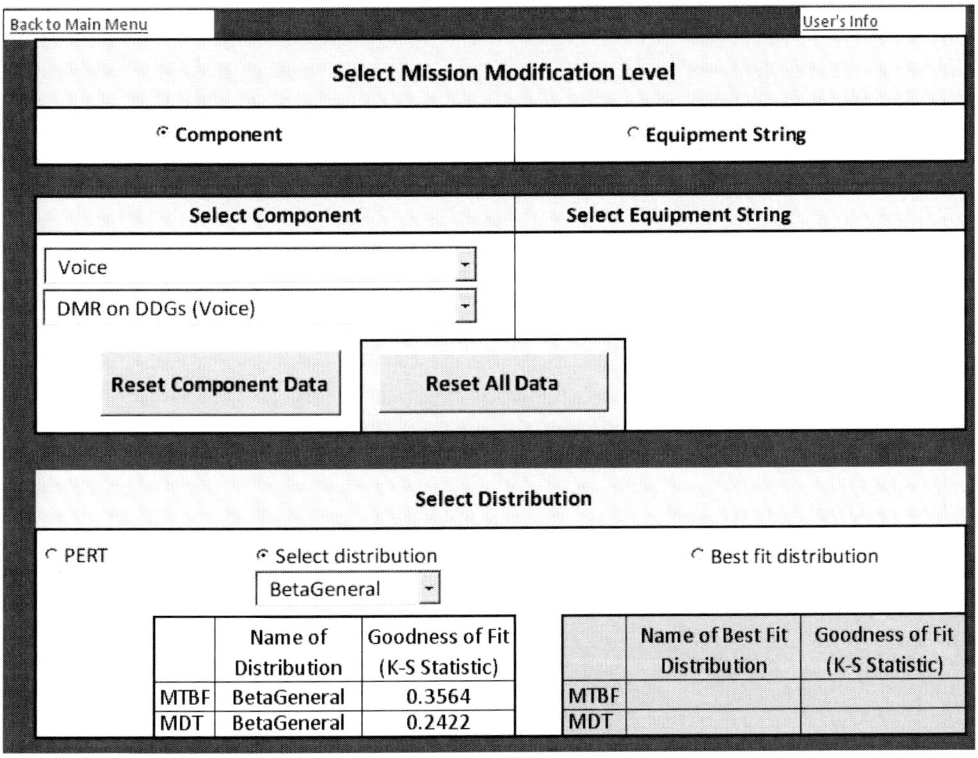

RAND *MG1003-6.4*

Figure 6.5
Front End of RAND's Tool in Component Mode: Selecting a Best-Fit Distribution

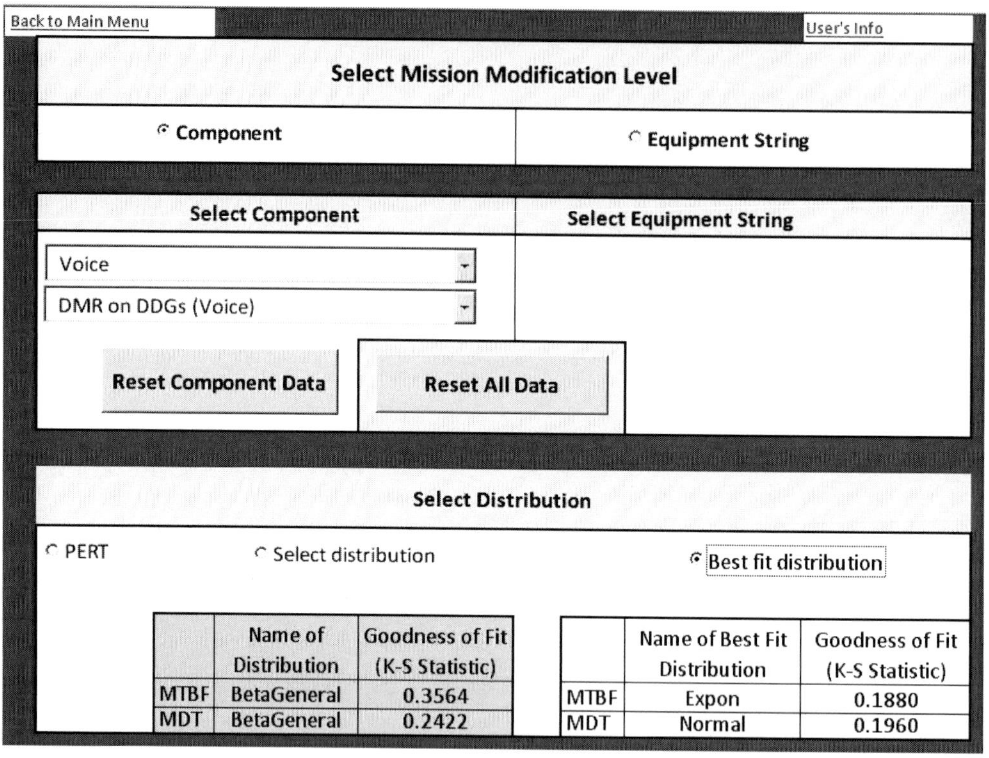

RAND *MG1003-6.5*

The Goodness-of-Fit Metric: The K-S Statistic

As seen in Figures 6.4 and 6.5, when either the "select distribution" or "best fit distribution" radio buttons are selected, the tool displays the chosen distribution as well as a goodness-of-fit metric for both MTBF and MDT. In our tool, we use the K-S statistic to measure this goodness of fit. While the chi-squared statistic is probably the most well-known goodness-of-fit metric, it is also dependent on how the sample data are binned. To avoid this drawback, we have chosen to use the K-S statistic instead. This statistic is a nonparametric test of equality between two one-dimensional distributions and is used to compare a sample with a reference distribution based on a form of minimum distance estimation. In particular, the distance between the sample and reference distribution is described by the parameter D_n, where

$$D_n = \sup_x | F_n(x) - F(x) |$$

and

- $F(x)$ is the reference cumulative distribution function

- $F_n(x) = \dfrac{N_x}{n}$

- N_x is the number of data points with value less than x

- n is the number of data points

- sup represents the supremum function, which is defined as the smallest real number that is greater than or equal to every number in the set. In other words, the sup of a set of real numbers is simply the least upper bound.

The smaller D_n is, the better the fit is of the data set to the distribution specified by $F(x)$. Hence, by definition, as shown in Figure 6.5, the best-fit distributions, namely exponential and normal for MTBF and MDT, respectively, have smaller K-S statistics than the manually selected Beta distributions for MTBF and MDT.

Operating RAND's Tool in "Equipment String" Mode

The RAND tool can also be operated in equipment string mode. In this mode, the user can modify the minimum, most likely, and maximum values of MTBFs and MDTs of all elements in the selected equipment string simultaneously. These modifications, of course, apply to the parameters of PERT distributions. If the user has previously opted to manually select a distribution or use a best-fit distribution for particular elements in the model, these modifications of PERT parameters will not affect prior user selections (as alternate distributions chosen by the user will be used in this case). For all elements in the model for which the user has not previously opted to manually select a distribution or use a best-fit distribution, a PERT distribution with the user-selected modifications will be used. Figure 6.6 shows a partial front-end view of RAND's tool in equipment string mode. In this example, the user has selected the "voice" equipment string, and next has the option of modifying the PERT parameters for all elements in this equipment string.

Outputs of RAND's Tool: Availablity Sensitivity Analyses

After the user has completed their component and system modifications of PERT parameters and selection of manually chosen or best-fit distributions for MTBFs and MDTs, they can run a Monte Carlo simulation that samples from all of these distributions. The simulation is run by a simple press of the "run simulation" button. The output of RAND's tool is an availability histogram and an availability sensitivity analysis tornado plot. We discuss each in turn below.

Figure 6.6
Front End of RAND's Tool: Equipment String Mode

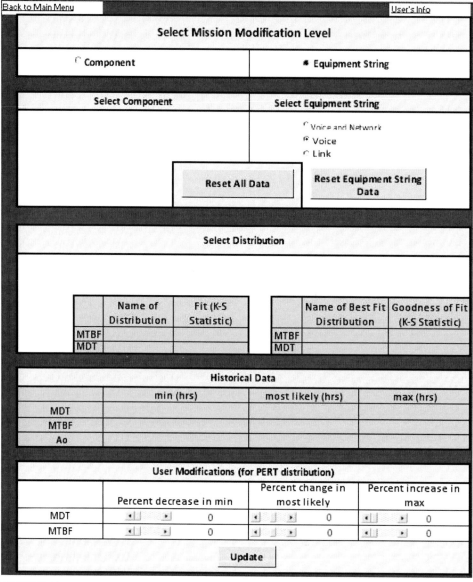

RAND *MG1003-6.6*

The Availability Histogram

The availability or Ao histogram is a probability density of the Monte Carlo results. Specifically, the y-axis values represent the relative frequency of a value in the range of a bin on the x-axis (number of observations in a bin/total number of observations) divided by the width of the bin. Figure 6.7 shows a partial front-end view of RAND's tool, depicting the availability histogram as well as the simulation settings the user has control of. Before running the simulation by pressing the corresponding button, the

Figure 6.7
Ao Histogram

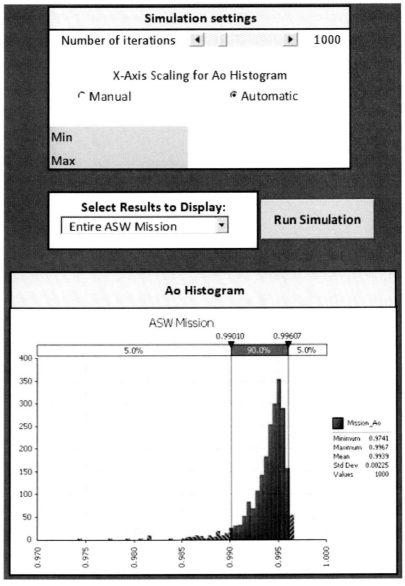

user has control over some of the simulation settings. First, the user has the option to set the number of Monte Carlo iterations, up to a maximum of 5,000, by adjusting a scroll bar. Next, the user may select between a manual or automatic x-axis scaling of the availability histogram. In manual mode, two scroll bars appear to adjust the minimum and maximum x-axis values. The user also has control over which output results they would like to view. Specifically, they may opt to view results for the entire ASW mission or for a particular equipment string. The user makes their selection from a pull-down menu under "Select Results to Display."

In the simulation depicted in Figure 6.7, all elements in the model with sufficient historical data used a best-fit distribution for both their MTBFs and MDTs, except for DMRs on DDGs, which used PERT distributions. All remaining elements used PERT distributions as well. The user chose to run 1,000 Monte Carlo iterations and to view results for the entire ASW mission. As shown in the statistics to the right of the plot, the availability histogram has a mean value for Ao of 0.9939 with a standard deviation of 0.00225. Statistics are also shown for the minimum and maximum values obtained. In addition, superimposed on the plot is the interval within which 90 percent of the values lie. Specifically, 90 percent of the availabilities obtained lie within 0.99010 and 0.99607.

The Availability Sensitivity Analysis

The availability sensitivity analysis shows a tornado plot of regression coefficients. (See Figure 6.8.) Specifically, it describes how a one-standard-deviation change in the mean values of the various inputs would impact the mean value of the output mission Ao, expressed as a factor of the output's standard deviation. The inputs are ranked in order from most to least impact on mission Ao. The bars to the left in the plot indicate that an increase in that element would cause a decrease in the output mission Ao, while the bars to the right in the plot indicate that an increase in that element would cause an increase in the output mission Ao.

Figure 6.8 shows the sensitivity analysis plot for the same simulation run as depicted in Figure 6.7. We see in Figure 6.8 that DMR on DDGs is the most critical component in terms of its effect on the overall mission availability. Specifically, a one-standard-deviation increase in the mean value of DMR's MTBF results in an increase in the mean value of the output mission Ao by a factor of 0.25 times its standard deviation.

Figure 6.8
Ao Sensitivity Analysis

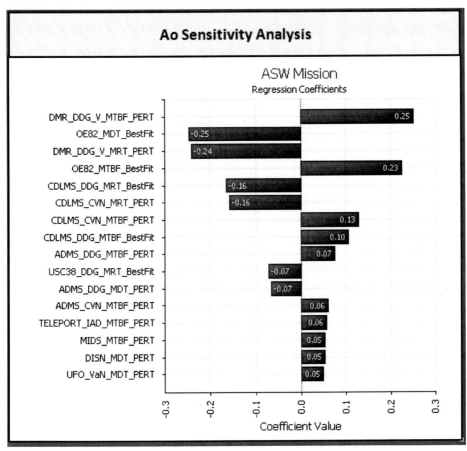

Summary of Key Features of RAND's Network Availability Modeling Tool

We outline the key features of RAND's network availability modeling tool below. The tool

- clearly shows key parameters of historical data on a per-component or per-system basis
- allows the user to modify these parameters
- allows the user to choose between a PERT distribution and, if sufficient data are available, a manually selected or best-fit distribution as well to model component and system MTBFs and MDTs

- allows the user to see the impact of modifying these parameters on mission or service (discussed in next chapter) availability
- allows the user to determine which components or systems in a mission or service have the greatest impact on the mission availability and further to visualize the magnitude of that impact
- has a simple graphical user interface that allows the user to easily navigate it without knowledge of Excel or @RISK.

This tool facilitates a more detailed accounting of hardware components and the uncertainty associated with such measurements. It also focuses consideration on performance of a service. It is thus offered as an improvement over the status quo approach. The new tool does not yet account for problems related to human error, human processes, and software, which can have a significant impact. Such an enhancement is recommended for future developments.

Comparing the New Tool with the Old Tool

We summarize here the results that compare the original SPAWAR 5.1.1 spreadsheet model (the old tool) with RAND's modifications to it (the new tool) for Ao values in the ASW mission's equipment strings. A major difference in results between the new model and the old one is that the new one is far less optimistic about component availability. This is due to the inclusion of real-world data and uncertainty. The major differences between the two efforts are enumerated as follows.

- The old model (SPAWAR 5.1.1 Pilot Effort):
  - used requirements' threshold and objective values for system and component performance specifications (MTBF and MDT)
  - employed a deterministic model (e.g., factored in no uncertainty) and thus generated a single value for equipment string and overall mission availability.
- The new model (RAND's Modified Model):
  - uses historical data (where available) instead of purely data from requirements documents
  - employs a stochastic modeling approach:
    - o fits historical data to probability distributions to describe system and component performance specifications (MTBF and MDT)
    - o generates distributions describing probability of a given availability for equipment strings and for the overall mission.

RAND's modified tool adds additional functionality to the old model as well. Specifically, the new tool performs sensitivity analysis of the systems and components in the model to determine their relative importance on the individual equipment strings and overall mission availability. For example, the new tool allows the user to isolate particular network segments, such as the ADNS and the ISNS, and to perform separate analyses on that portion of the network.

Figure 6.9 shows a large range of potential availabilities. This is because the new tool accounts for uncertainty in measures that drive availability. Ninety percent confidence intervals are shown on the plots in darker blue. Means are shown as white dots.

Figure 6.9
Comparison of Mission Availability: Old Versus New Model

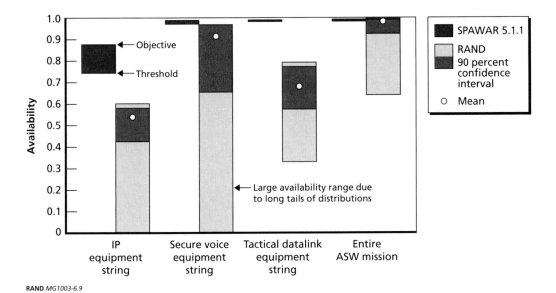

CHAPTER SEVEN

Exemplar Analysis Using the New Tool

This chapter provides analysis using the tool described in the previous chapter. Specifically, we calculate the availability histograms and the sensitivity analysis tornado plots for both the ASW mission and the SameTime chat and COP services when best-fit distributions are used (i.e., when sufficient data exist). For those components and systems with limited or no historical data, we use PERT distributions. For the chat and COP service models, we obtained historical data for KG-175, KG-84, MD-1324 Modem, WSC-6, WSC-6 Terminal, WSC-8, and WSC-8 Terminal. For the remainder of the components in the chat and COP service models, we used manufacturer's specifications as the most likely value in a PERT distribution and plus and minus 20 percent of this value for the maximum and minimum values, respectively. The list of components for which we obtained historical data as well as the list of components for which manufacturer's specifications were used are summarized in Chapter Four.

Analysis of the ASW Mission

Figure 7.1 shows the availability histogram for the ASW mission. We see that ASW has a mean availability of 0.9838, with a standard deviation of 0.0397. Ninety percent of the values lie within 0.9345 and 0.9961.

Figure 7.2 is the sensitivity analysis tornado plot, which shows that DMR on DDGs is the most critical system in terms of its impact on the mission availability. Specifically, a one-standard-deviation increase in the MDT (referred to as MRT [mean restoral time] in the plot) for DMR on DDGs, which translates to an increase from 218 hours to 309 hours (42 percent increase), results in a decrease in the mean value of the output mission Ao by a factor of −0.41 times its standard deviation, or a reduction by 2 percent, from 0.9838 to 0.9675. A one-standard-deviation increase in the MTBF of DMR has a much smaller effect on the mission availability, namely, an increase by a factor of 0.10 times its standard deviation. The next most critical components in the ASW mission are the MDTs of MIDS, INTELSAT, and CDLMS on CVNs.

Next, we analyze each of the three parallel equipment strings in the ASW mission—IP network, secure voice, and tactical datalink—separately.

Figure 7.1
Availability Histogram for the ASW Mission

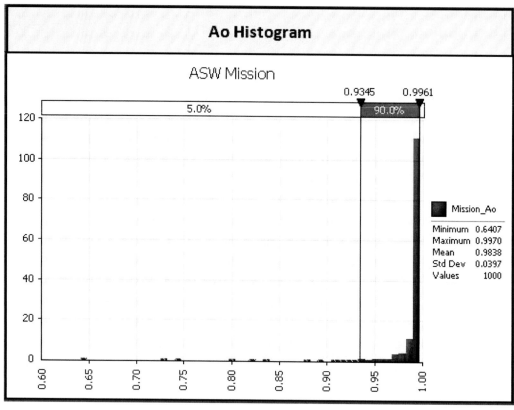

Figure 7.2
Sensitivity Analysis for the ASW Mission

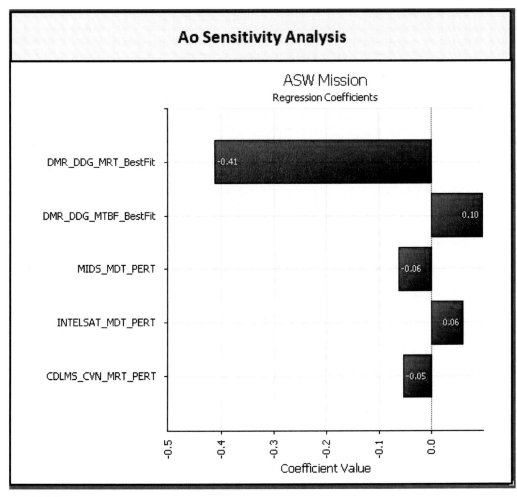

Analysis of the IP Network Equipment String

Figure 7.3 shows the availability histogram for the IP network equipment string. We see that this equipment string has a mean availability of 0.5417, with a standard deviation of 0.0780. Ninety percent of the values lie within 0.416 and 0.588.

Figure 7.4 is the sensitivity analysis tornado plot, which shows that DMR on DDGs is the most critical system in terms of its impact on the availability of the equipment string. Specifically, a one-standard-deviation increase in the MDT (referred to as MRT in the plot) for DMR on DDGs results in a decrease in the mean value of the equipment string availability by a factor of −0.42 times its standard deviation, or a reduction by 6 percent, from 0.5417 to 0.5089. The next most critical elements in the IP network equipment string are the MDT and MTBF of ADMS on DDGs, and the MTBF of DMR on DDGs.

Figure 7.3
Availability Histogram for the IP Network Equipment String

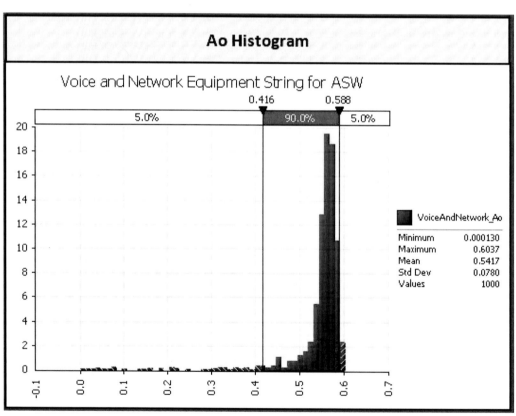

Figure 7.4
Sensitivity Analysis for the Voice and Network Equipment String

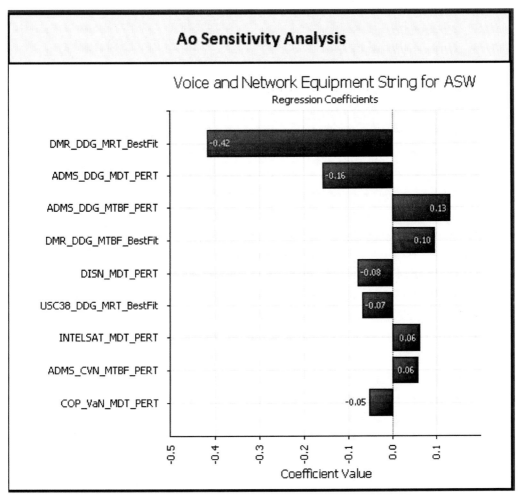

Analysis of the Secure Voice Equipment String

Figure 7.5 shows the availability histogram for the secure voice equipment string. We see that this equipment string has a mean availability of 0.9128, with a standard deviation of 0.1460. Ninety percent of the values lie within 0.648 and 0.967.

Figure 7.6 is the sensitivity analysis tornado plot, which shows that DMR on DDGs is the most critical system in terms of its impact on the availability of the equipment string. Specifically, a one-standard-deviation increase in the MDT (referred to as MRT in the plot) for DMR on DDGs results in a decrease in the mean value of the equipment string availability by a factor of −0.40 times its standard deviation, or a reduction by 6 percent, from 0.9218 to 0.8634. The next most critical element in the secure voice equipment string is the MTBF of DMR on DDGs.

Figure 7.5
Availability Histogram for the Secure Voice Equipment String

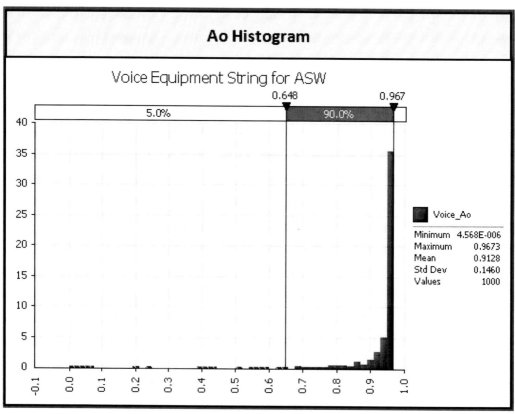

Figure 7.6
Sensitivity Analysis for the Secure Voice Equipment String

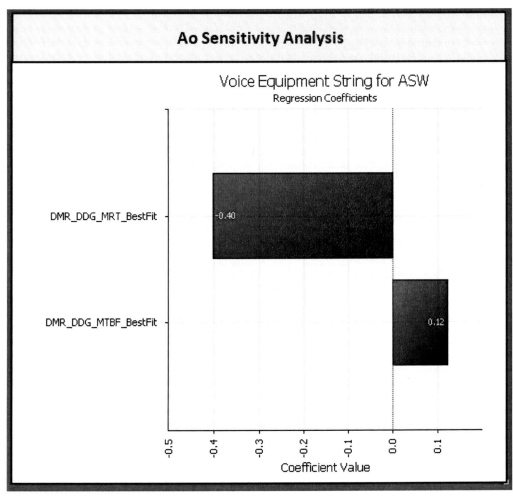

Analysis of the Tactical Datalink Equipment String

Figure 7.7 shows the availability histogram for the tactical datalink equipment string. We see that this equipment string has a mean availability of 0.6807, with a standard deviation of 0.0546. Ninety percent of the values lie within 0.5797 and 0.7523.

Figure 7.8 is the sensitivity analysis tornado plot, which shows that CDLMS on DDGs is the most critical system in terms of its impact on the availability of the equipment string. Specifically, a one-standard-deviation increase in the MDT (referred to as MRT in the plot) for CDLMS on DDGs, which translates to an increase from 601 hours to 965 hours, results in a decrease in the mean value of the equipment string availability by a factor of –0.49 times its standard deviation, or a reduction by 4 percent, from 0.6807 to 0.6539. The next most critical elements in the tactical datalink equipment string are the MDT of CDLMS on CVNs, the MTBF of CDLMS on CVNs, and the MTBF of CDLMS on DDGs.

Figure 7.7
Availability Histogram for the Tactical Datalink Equipment String

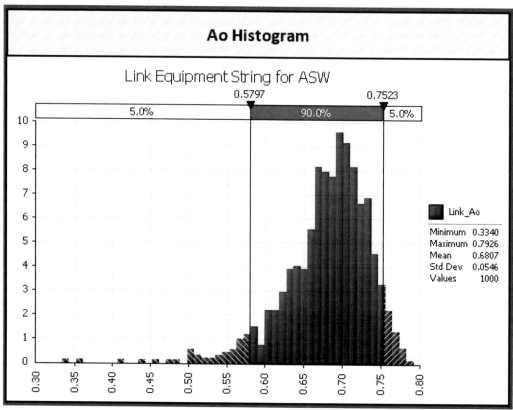

Figure 7.8
Sensitivity Analysis for the Tactical Datalink Equipment String

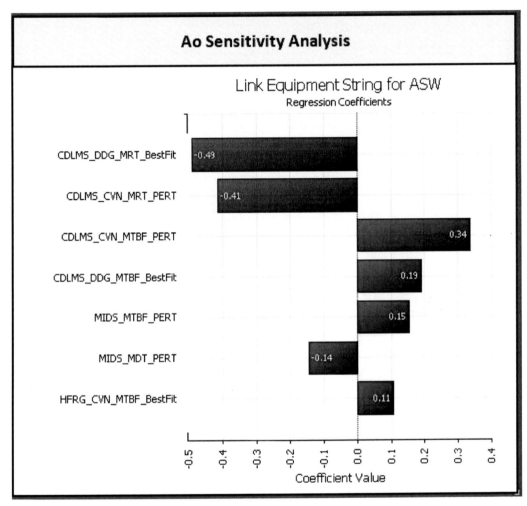

Summary of Results for ASW Mission

In Tables 7.1–7.4, we summarize those components or systems with the top three (in magnitude) regression coefficients for the entire ASW mission as well as for each of the three individual equipment strings. We find that the critical systems are DMR, ADMS, and CDLMS.

Table 7.1
ASW Mission Regression Coefficients

System or Component	Regression Coefficient
MDT of DMR on DDGs	−0.41
MTBF of DMR on DDGs	+0.10
MDT of MIDS	−0.06

Table 7.2
IP Network Equipment String Regression Coefficients

System	Regression Coefficient
MDT of DMR on DDGs	−0.42
MDT of ADMS on DDGs	−0.16
MTBF of ADMS on DDGs	+0.13

Table 7.3
Secure Voice Equipment String Regression Coefficients

System	Regression Coefficient
MDT of DMR on DDGs	−0.40
MTBF of DMR on DDGs	+0.12

Table 7.4
Tactical Datalink Equipment String Regression Coefficients

System	Regression Coefficient
MDT of CDLMS on DDGs	−0.49
MDT of CDLMS on CVNs	−0.41
MTBF of CDLMS on CVNs	+0.34

Analysis of the Chat Service

Figure 7.9 shows the availability histogram for the chat service. We see that this service has a mean availability of 0.99269, with a standard deviation of 0.000395. Ninety percent of the values lie within 0.992007 and 0.993338.

Figure 7.10 is the sensitivity analysis tornado plot, which shows that the most critical component in terms of its impact on the availability of the chat service is TRANSEC KG-194. Specifically, a one-standard-deviation increase in the MTBF for TRANSEC KG-194 results in an increase in the mean value of the chat service availability by a factor of 0.49 times its standard deviation. A one-standard-deviation increase in the MDT for TRANSEC KG-194 results in a decrease in the mean value of the chat service availability by a factor of –0.48 times its standard deviation. The next most critical elements in the chat service are MIDAS and the Nextera Multiplexer. We list their regression coefficients in Table 7.5. We note here that these results are dependent

Figure 7.9
Availability Histogram for the Chat Service

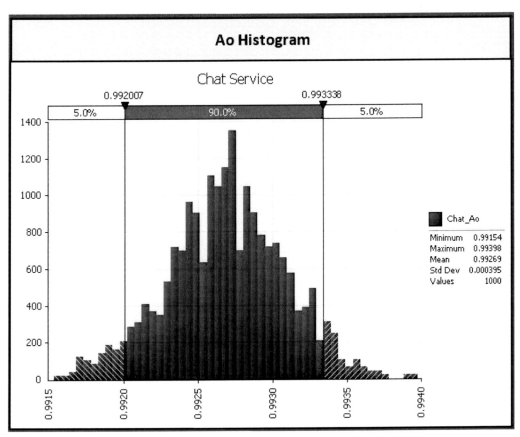

Figure 7.10
Sensitivity Analysis for the Chat Service

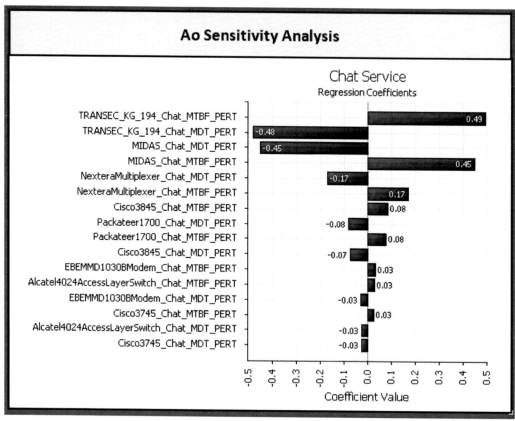

on our assumptions of PERT distributions to describe the MTBFs and MDTs of those components with limited or no historical data. In addition, the results are also dependent on the widths of those PERT distributions. As mentioned earlier, we have chosen to use plus and minus 20 percent of the manufacturer's specification as the maximum and minimum values of the corresponding PERT distribution.

Table 7.5
Chat Service Regression Coefficients

Component	Regression Coefficient
MTBF/MDT of TRANSEC KG-194	+0.49/−0.48
MTBF/MDT of MIDAS	+0.45/−0.45
MTBF/MDT of Nextera Multiplexer	+0.17/−0.17

Analysis of the COP Service

Figure 7.11 shows the availability histogram for the COP Service. We see that this service has a mean availability of 0.99252, with a standard deviation of 0.000395. Ninety percent of the values lie within 0.991841 and 0.993169.

Figure 7.11
Availability Histogram for the COP Service

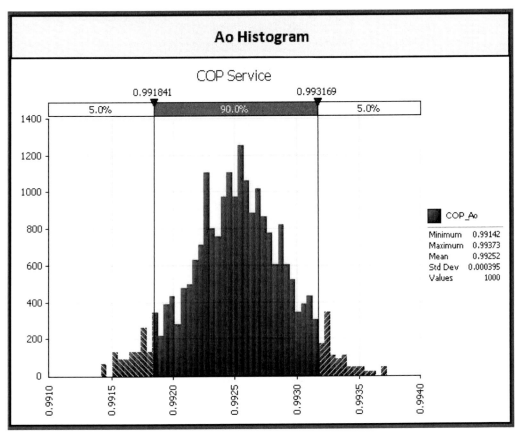

Figure 7.12 is the sensitivity analysis tornado plot for the COP service. The results are quite similar to those seen for the chat service. The sensitivity analysis shows that the most critical component in terms of its impact on the availability of the COP service is TRANSEC KG-194. Specifically, a one-standard-deviation increase in the MTBF for TRANSEC KG-194 results in an increase in the mean value of the COP service availability by a factor of 0.49 times its standard deviation. A one-standard-

Figure 7.12
Sensitivity Analysis for the COP Service

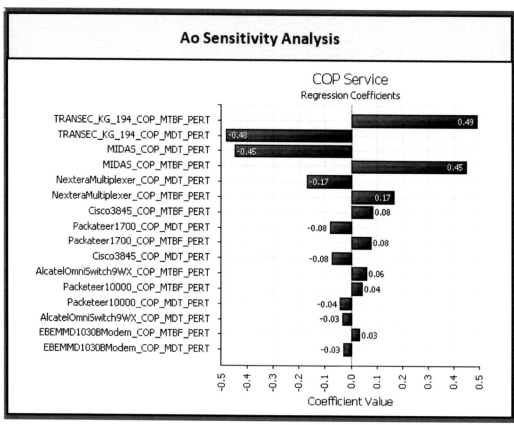

deviation increase in the MDT for TRANSEC KG-194 results in a decrease in the mean value of the COP service availability by a factor of –0.48 times its standard deviation. The next most critical elements in the COP service are MIDAS and the Nextera Multiplexer. We list their regression coefficients in Table 7.6.

Table 7.6
COP Service Regression Coefficients

Component	Regression Coefficient
MTBF/MDT of TRANSEC KG-194	+0.49/–0.48
MTBF/MDT of MIDAS	+0.45/–0.45
MTBF/MDT of Nextera Multiplexer	+0.17/–0.17

CHAPTER EIGHT

Conclusions, Recommendations, and Next Steps

Conclusions

We conclude the following:

1. Precise models of dependability, and of availability in particular, are difficult to create: The correlations between the human user, hardware, and software are difficult to capture, as is the composition of services.
2. A disconnect between fleet-measured availability and user experiences is conceivable. Fleet data could suggest that availability is high when users experience it as low if all the components involved in the delivery of a user service are not fully accounted for and if other sources of error (e.g., software) are not included in the analysis. Fleet data tend to focus on specific *equipment* failures; users perceive poor *service*, which is impacted by many factors, including the users' actions, software, and other environmental factors.
3. The most meaningful metrics for gauging Navy network performance—from a user perspective—are ones tied to performance of a service. Measurements of user experiences should be synthesized from qualitative surveys (trouble ticket reports, help desk feedback, etc.).

One overarching conclusion of this study is that operational availability, Ao, as defined by OPNAV Instruction 3000.12A, is an outdated measure that is too general and does not reflect users' experience with respect to IT and network services. It is worth noting that one of the driving forces for this study was the conflicting reports received by PEO C4I regarding the Ao of C4I networks—engineers would report close to 100 percent availability, whereas users would consistently complain about network outages. The reason for this discrepancy lies in the way Ao is measured. The OPNAV definition is more suitable for hardware. However, hardware-only availability calculations[1] for networks are inherently uncertain. A problem with Ao as a metric is that it is

---

[1]   Also, NSWC-Corona reliability block diagrams are hardware-oriented and do not always take into account software applications.

primarily intended for defining requirements' specifications for equipment acquisition, and not for operational analysis. Therefore, there are times when a network may be operating neither in an up nor a down state, but rather in a degraded state, such that some users may perceive it as operational but others may not.[2]

Perhaps the definition of *downtime* for availability calculations should be redefined to include "gray areas" where a system is functional (or "up") but below user expectations; perhaps availability should be specified for individual user services. This requires additional study, and a precise recommendation is beyond the scope of this report.

In this study, we do not propose a complete departure from the hardware-based analysis done currently—at least not immediately. This is because much of the currently accessible data from the fleet document hardware performance (that can be easily observed to be either up or down). As a result, we utilize the hardware-based availability data that exist, but we can mitigate the weakness of a solely hardware-oriented analysis by (1) the incorporation of uncertainty and (2) a focus on services as perceived by the user.[3] Nonetheless, there is a need to capture and report human impacts on the network and software/application impacts on systems and services. This will enable more-sophisticated and more-accurate models of network dependability.

Recommendations

We recommend a number of steps that the PEO C4I might take to facilitate future network dependability assessments, provide more meaningful results to network engineers, and, ultimately, enhance the dependability of C4I networks across the fleet.

Create a Single Accessible Portal for Network Diagrams

A major challenge in this study was the gathering and consolidation of disparate sources of data for network architectures. A single data repository for these diagrams would facilitate not only reliability studies but also provide the network engineers with a holistic, bird's-eye view of afloat networks. Anecdotally, one of the most common remarks heard during our interviews was, "There is no one in the Navy with a comprehensive view of our network infrastructure. We need such a person." The creation of this portal would empower such personnel and encourage them to create holistic solutions for the entire network, rather than focusing their talents on their individual enclaves.

---

[2]   Traffic congestion is a common cause of such behavior.

[3]   This approach takes into account the various dependencies associated with a given service. For instance, in the case of SameTime chat, the DNS server is critical to successful operation and is included in the service-oriented model.

The Navy's creation of a website intended specifically for ISEAs is a positive step in this direction, but, as of this writing, the site still does not contain shore network diagrams. Since most afloat services are dependent on the data in shore servers, the addition of the shore infrastructure would greatly enhance the usability of this site.

Pursue User-Perceived Service and Mission Availability Metrics

User-perceived service availability is the percentage of service invocations requested by a particular user that are successful (out of the total number of service invocations attempted by that user) during a given time interval. We believe that user-perceived service availability is the most appropriate metric for Navy networks because network availability from the user's perspective is more valuable than service availability data measured by an automated instrument.

We can extend the scope of this metric to describe the effect of network reliability on a given mission. We use the term *mission availability* to encompass not just the availability of a given service during the course of a mission, but also the impact that the service's availability has on the mission. Specifically, we define mission availability as

$$A_{mission} = \sum_{j=1}^{M} w_j \times \sum_{i=1}^{N} A_{i,j} u_{i,j}$$

where $A_{mission}$ is the mission availability, M is the number of services that have a mission impact, N is the number of users, w_j is the weight of service j on the impact of the mission, and $A_{i,j}$ and $u_{i,j}$ are system availabilities and usage factors, respectively, for user i and service j.

Be Cognizant of the Need to Design Sufficient Human-Machine-Interfaces

Flawed human processes and corresponding human errors are to blame for many network outages. As noted in interviews conducted by Schank et al. (2009),

> [S]ailors do not maintain the equipment well. The equipment is complicated and the sailors are not trained and they do not read the manuals. Now that manpower is being reduced so rapidly it leads to the problem of there being no one to maintain the equipment.

Proper training and human-engineering design considerations could lower the number of outages caused by human operators. Information Technology Readiness Review data from the USS *Eisenhower*, USS *Lincoln*, and USS *Halyburton* reveal that many casualty reports specified human error, not equipment failure, as the root cause. Network reliability models should factor this in to be more accurate. In fact, in a study of DoD human-engineering methods, Pat O'Brien (2007) points out that the human can be seen as a "black box" in the system with its own coefficients of availability and reliability. In the case where a human is not strong enough to lift a chassis into place, insert or eject a part, the human can be seen as being unavailable, leading to the system being unrepairable. In the case of the interface being confusing, the human is driven to perform his/her duties unreliably.

A common thread in the literature (see Yamamura et al., 1989; Maxion and Reeder, 2005; Reeder and Maxion, 2005) on interface design is the direct relationship between human error and poorly designed human-machine interaction (HMI) interfaces. Such studies can guide the Navy in its pursuit of more uniform interfaces[4,] for its afloat IT staff.[5] The Navy should consider creating service-level agreements with clear definitions of acceptable HMI designs.

Next Steps

The most valuable application of network dependability metrics for the Navy is to measure the impact of network dependability on mission success. To accomplish this, future work should incorporate the ability to analyze the impact of network dependability on ship and multiship operational capabilities. This could be integrated into the new tool described in this report. Furthermore, we propose building on this study's proposed framework to address human error. This analysis would require a more rigorous data-gathering effort, with the end goal being to fully accommodate the impact of the human element on component and/or system availability and reliability.

Find Ways to Fuse Multiple Data Sources for Network Availability

During the course of this study, we encountered numerous sources of availability and reliability data. NSWC-Corona is the most visible source of Reliability, Maintainability, and Availability (RMA) data, and they served as our primary source for component data in creating our models. ISNS and ADNS trouble-ticket data from Remedy provided the basis of our human-error calculations. There were other data sources as

---

[4] For instance, ADNS Increment II saw a migration to Cisco hardware and software as the standard for networking equipment. To an extent, this creates a uniform interface for network engineers and thus contribute to the reduction of human error with respect to the Cisco interface.

[5] It is worth noting that, to address the problem of human error in terms of network dependability, commercial giant Motorola turned to DoD human-engineering methods. According to O'Brien (2007), Motorola intends to follow Mil-HDBK-46855A Human Engineering Program Process and Procedures (U.S. Department of Defense, 2006) to reduce procedural errors. Motorola defined the problem as one of reducing procedural or human errors when the root cause was not in the written procedures themselves but rather in the standards or platform design. Although several standards have tried to define HMI procedural errors, Motorola discovered that such standards were incomplete and did not alleviate what it found to be a "disturbing" industry-average 50-percent downtime caused by such errors. The result of Motorola's efforts was the publication of multiple human-engineering design reviews and studies relevant to its network infrastructure (O'Brien, 2007).

well that provided valuable insight into network reliability. However, due to their varying formats, they did not lend themselves to a standardized data-extraction process. In addition, we learned that Regional Maintenance Centers (RMCs) and port engineers may have their own trouble ticket systems or databases that would further add to the current stock of availability data. The way ahead probably requires the creation of a new database to be populated manually from these reports to make them usable.

Tie Mission Simulation Tools into Availability Calculation

A next step should also be to add the capability to simulate the operational impact of network dependability. This analysis will help determine the relationship between availability and mission success by tying in mission-thread simulators. This analysis can also help account for geographical effects and thus quantify how much availability is needed to achieve mission success.

Accommodate the Impact of the Human Element

A framework for accommodating the human factor into our tool is needed. A complete service model must consider the human process and human user factors that can affect the downtime of individual components, such as level of training and the frequency of human interactions with a particular component.

Enhance the New Tool: Make It Web-Based and Automated

As suggested by the PMW 750 CVN C4I CASREP study report (Conwell and Kolackovsky, 2009), PEO C4I has a need to establish a C4I performance dashboard using Corona data. The framework and tool described in this research can contribute to that goal. Toward this goal, RAND recommends that the newly developed network availability modeling tool described herein be made web-based and upgraded to automatically incorporate the latest historical data from NSWC-Corona as they are made available.

A web-based tool would be more easily and widely distributed across the Navy, and would typically yield more usage. The front-end interface for the tool should be intuitive and require minimal training to use.

The tool should also have an external data interface, allowing for automatic updates of data. We would need to define a data-exchange standard, such as XML, so that the tool can more easily assimilate updated data and incorporate new sources of data. These modifications would enable us to fuse other data sources as described above and to incorporate mission impacts to allow more relevant sensitivity analysis. Figure 8.1 shows a sample architecture for such an automated web-based tool.

Address User Expectations

There is some speculation that many network users expect more than the system is designed to deliver. Some percentage of user dissatisfaction could be excessive expec-

Figure 8.1
Sample Architecture for Automated, Web-Based Tool

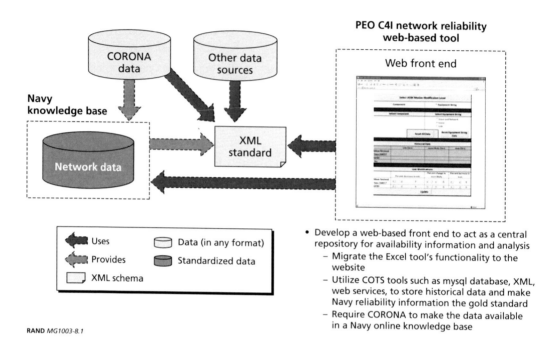

RAND *MG1003-8.1*

tation. For example, it is conceivable that on ships with "small pipes," this under-provisioning can be a bigger factor than system malfunction. In summary, the user perspective could suggest system performance that is poorer than any accurately determined metric value. It may be worthwhile to attempt to quantify this possibility of "false" perception in future work.

RAND Review of Initial SPAWAR Tool

RAND Tasking

SPAWAR 5.1.1 conducted a pilot study on the End-to-End Operational Availability for Antisubmarine Warfare (ASW) Missions. This initial effort built a hard-coded spreadsheet that calculated network equipment availabilities. A screen shot is shown in Figure A.1.

RAND reviewed this tool and the accompanying written report (Larisch and Ziegler, 2008a). Overall, RAND provided mathematical corrections to the SPAWAR report, and built on the existing spreadsheet by incorporating probabilistic models of component availability.

Figure A.1
Screen Shot of SPAWAR Tool

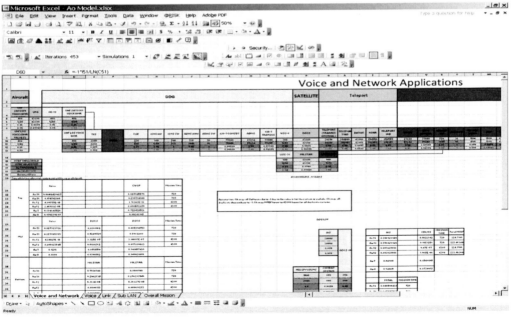

RAND *MG1003-A.1*

SPAWAR Approach

Data

SPAWAR used requirements data to populate its spreadsheet, and RAND adopted ranges of Ao values to define ranges of uncertainty, using data from NSWC-Corona to bound these uncertainties. RAND then performed sensitivity analyses via Monte Carlo methods.

The SPAWAR 5.1.1 study used a constant value of 2.5 hours for mean restoral time (MRT) for every component in its model. To understand the ramifications of this broad assumption upon Ao, we simulated a range of MRT values and examined the resulting mission Ao values. Figure A.2 shows the resulting availabilities for both 30-day and 6-month missions for MRT values ranging from 1.5 to 3.5 hours. This figure was generated by using the existing SPAWAR model and varying only the deterministic value for MRT.

Figure A.2 demonstrates how minor variations in MRT lead to greater than 1 percent fluctuations in Ao. Although this may seem like an inconsequential change, consider that Ao requirements are often made up to three significant figures—for instance, the industry standard is "five nines," or 99.999 percent. If a global MRT change of two hours decreases the overall Ao of a given system from 99.999 percent to 98.999 percent, the design clearly needs to be overhauled to meet system specifications.

In further iterations, we modified the mean value of MRT per equipment string to be 3 hours, instead of the previous mean value of 2.5, keeping the range at plus and minus 10 percent of the mean. Figure A.3 shows the resulting relative frequency his-

Figure A.2
Sensitivity to MRT

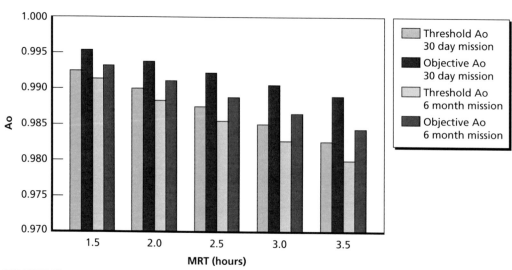

Figure A.3
Threshold Availability

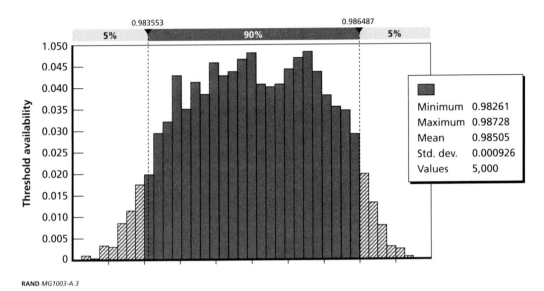

togram for end-to-end overall mission threshold availability. We note that reliability is not impacted by MRT, and hence we do not plot an additional histogram for it here. As expected, when the mean value of MRT was increased from 2.5 to 3 hours, the mean value for overall availability decreased. Specifically, it decreased from 0.98751 to 0.98505.

Tool

SPAWAR 5.1.1 built an Excel spreadsheet to calculate the availability Ao and reliability values for each equipment string and for the overall mission in the ASW model, using RMA requirements data from PEO C4I Programs of Record (POR). SPAWAR developed two versions of equipment strings: a component-level version and a POR-level version. The component-level string was derived from a high-level operation view (OV-1) describing the ASW mission thread, operational activity sequence description (OV-6c) describing the sequence of activities involved in the mission, operational information exchange matrix (OV-3) capturing the dependencies among OV-6c activities, the PEO C4I Master Plan for system architecture, and the PEO C4I PR09 Integrated Roadmap to identify the PORs providing the various functionalities to the fleet. The Excel spreadsheet took the resulting equipment strings from this effort and aggregated them up to the POR level.

Availability Calculation

SPAWAR 5.1.1 used two different approaches to calculate end-to-end Ao. In the first approach, they used Ao's specified per individual component to arrive at an end-to-end overall mission Ao. Component-level Ao's were combined, depending on whether the components are in series or parallel, as follows:

$$Ao_{E2E\text{-serial}} = Ao_1 \times Ao_2 \times Ao_3 \ldots$$

$$Ao_{E2E\text{-parallel}} = 1 - [(1 - Ao_1) \times (1 - Ao_2) \times (1 - Ao_3) \times \ldots]$$

In the second approach, SPAWAR 5.1.1 used individual component-level MTBF data to ultimately arrive at an end-to-end overall mission Ao. MTBF is related to Ao according to the standard definition provided in OPNAV Instruction 3000.12A, namely

$$Ao = \frac{MTBF}{MTBF + MTTR + MLDT} = \frac{MTBF}{MTBF + MRT},$$

where MRT (mean restoral time) = MTTR (mean time to repair) + MLDT (mean logistics delay time).

Before arriving at an end-to-end overall mission Ao, however, the MTBF data for each component were translated into a component reliability, as follows:

$$R(t) = e^{-t/MTBF},$$

where t is mission duration. Serial and parallel strings for reliability were then calculated in a similar manner to that shown above for availabilities, namely:

$$R_{E2E\text{-serial}} = R_1 \times R_2 \times R_3 \ldots$$

$$R_{E2E\text{-parallel}} = 1 - [(1 - R_1) \times (1 - R_2) \times (1 - R_3) \times \ldots].$$

Finally, the end-to-end overall mission MTBF was calculated as

$$MTBF_{E2E} = \frac{-t}{\ln(R_{E2E})}.$$

Since MTTR and MLDT data were not available in the POR acquisition documentation, the study instead used assumed values for both, as encapsulated by a single assumed value for MRT of 2.5 hours. Using the assumed data for MRT and the end-to-end value for MTBF as calculated above, the study determined end-to-end Ao as the end-to-end MTBF divided by the sum of the end-to-end MTBF and MRT.

RAND Recommendations

RAND recommended the following corrections and improvements to the SPAWAR 5.1.1 study:

1. Correct minor calculation errors and oversights.
2. Incrementally expand the resolution of the model.
3. Use nondeterministic values (MTBF, MTTR, MLDT) from fleet data instead of fixed requirement values.
4. Wherever data (MTBF, MTTR, MLDT) are missing or unavailable, parameterize these values in the model.

Correct Minor Calculation Errors and Oversights

RAND discovered the following errors in the SPAWAR 5.1.1 spreadsheet model:

- For the "Voice and Network" tab:
  1. C30, C31, and C32 formulas are incorrect. They include "TVS" (E9) in "Parallel Voice" (C43), but model already puts "TVS" in "DDG Serial" (D43).
  2. Cells G26 and G27 are incorrect. KIV-7 TRANSEC (AN4) is left out of availability calculations.
  3. TELEPORT IAD (AN9) was left out of reliability and availability calculations.
  4. KIV-7 COMSEC (AT9) was left out of availability calculations for CVN serial (H43).

- For the "Voice" tab:
  1. C21 includes components that it should not.

- For the "Link" tab:
  1. A handful of the cell formulas used in the spreadsheet model were hard-coded instead of linked back to their appropriate data cell.

Incrementally Expand the Resolution of the Model

Component-level aggregation to POR-level is an oversimplification that does not provide useful means of identifying and rectifying bottlenecks. We understand the need to aggregate to a POR-level for the sake of expediency, but doing so eliminates the granularity necessary to identify points of failure.

For instance, the entire ADNS architecture is condensed into a single Excel column. This does not do justice to the rich architecture of ADNS and does not provide the analyst with the ability to identify a clear bottleneck. In this particular example, ADNS is the cornerstone of satellite communications, and therefore, merits more detailed modeling. A high-level architectural diagram of ADNS Increment II for CG and DDG platforms yields more than ten separate components, each of which is criti-cal to the success of a given communication link, as illustrated in Figure A.4.

In Figure A.4, the components enclosed in the red box are collapsed in the SPAWAR 5.1.1 spreadsheet into two cells—ISNS HW and ISNS SW. Those compo-nents in the blue box are collapsed into two other cells—ADNS HW and ADNS SW. As the figure demonstrates, this results in an oversimplification of the network archi-tecture and eliminates the granularity that is needed for a useful reliability analysis, especially with regards to the identification of bottleneck components.

Figure A.4
Abstraction Issue Example: ADNS Force Level Increment II

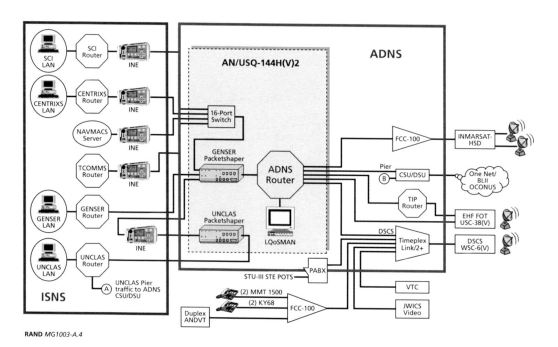

Detailed Analysis of Human Error in CVN Networks

The main body of this report described the many factors that affect network dependability, including the human impact. One parameter of the human impact on network dependability is the training of the individuals. In this appendix, we examined data on how training impacts the human effect on network dependability.

Specifically, in this appendix, we considered the human's role in network dependability (e.g., reliability) for a very specific data set. We examined the extent to which the numbers of personnel with the required technical training, among a ship's IT crew, affects the number of problems reported about the ISNS via the RTTS.

This examination was done in two parts. First, we examined the different levels of trained crew on all ten U.S. carriers in service for a three-year period starting in 2006. Second, we quantified the network reliability for those particular ships by calculating the number of human-related technical problems encountered. The data we used were from ISNS RTTS data from 2006 to 2008. For our purposes, the terms *trained crew*, *trained sailors*, or *trained personnel* refer to those individuals qualified for the ISNS.

More data and more analysis are needed to draw any firm conclusions. Nonetheless, we can say that future models of human impacts (on network dependability) could result from further analysis of data of this type. The main objective of this appendix is to show an example of how such data may be collected. It can help develop network dependability models that account for human induced errors.

ISNS Remedy Trouble Ticket Data

The RTTS data we reviewed were for the ISNS on ten carriers over three separate years. They are itemized in Table B.1.

The last column in Table B.1 lists the total number of trouble tickets attributed to human error. This was done as follows. The trouble ticket reports that were classified under "configuration" and "administrator" categories were assessed to be human-

Table B.1
Remedy Trouble Ticket Data

Ship	Trained IT Staff Onboard	2006		2007		2008		Total
		# of Trouble Tickets	# of Tickets Attributed to Human Error	# of Trouble Tickets	# of Tickets Attributed to Human Error	# of Trouble Tickets	# of Tickets Attributed to Human Error	# of Tickets Attributed to Human Error (2006–2008)
CVN 71 Theodore Roosevelt	0	232	7	96	42	74	39	88
CVN 73 George Washington	1	35	15	49	17	33	17	49
CVN 68 Nimitz	2	73	19	40	27	19	11	57
CVN 76 Ronald Reagan	2	192	12	111	16	16	13	41
CVN 65 Enterprise	3	150	46	113	25	13	9	80
CVN 69 Dwight D. Eisenhower	3	194	35	140	15	55	21	71
CVN 75 Harry S. Truman	5	51	8	59	21	18	5	34
CVN 70 Carl Vinson	7	1	0	1	0	14	2	2
CVN 74 John C. Stennis	8	69	14	19	6	19	12	32
CVN 72 Abraham Lincoln	10	353	15	73	17	44	3	35

related based on the description as well as discussions with ISNS experts.[1] It is noteworthy that we found that human-error trouble tickets were mostly due to software problems, specifically configuration and administrator subcategories. We also note that the percentage of all trouble tickets attributed to human causes increased over the years: 12 percent in 2006, 26 percent in 2007, and 43 percent in 2008. Overall, our examination of the data in Table B.1 suggests a pattern: The carriers with more of the billets filled with trained personnel experience lower numbers of trouble tickets overall and lower numbers of trouble tickets due to human error.

Discussion

Our initial observation is that the ships that accounted for most of the human errors across the whole group (of CVNs) usually had the lowest number of trained personnel (see Table B.2). More study is needed.

Table B.2
Expected Number of Human Error Trouble Tickets Compared with Number of Trained Sailors per Carrier

Number of Trained Sailors Per Carrier	Expected Number of Human Error Trouble Tickets Over Three Years, per Carrier
0	71
1	65
2	60
3	55
4	49
5	44
6	39
7	33
8	28
9	23
10	17

---

[1]   The process we used to determine human error was as follows:

1.   Eliminated all hardware issues from spreadsheet.
2.   Grouped all software-related errors that lacked descriptions into one category.
3.   Took software cases with unmistakable human error and noted their Category/Type/Item (CTI) categorization.
4.   Eliminated all software errors from #2 that did not match identical CTI categorization as #3.
5.   Categorized all configuration issues as human error based on random sampling of about 50 cases that we confirmed were all human-error related.

Network Diagram of the COP Service

Figures C.1 through C.3 show our network diagram of the COP service. In these diagrams, dual backbone switch architecture was not taken into account.

Figure C.1
Network Diagram of the COP Service (1 of 3)

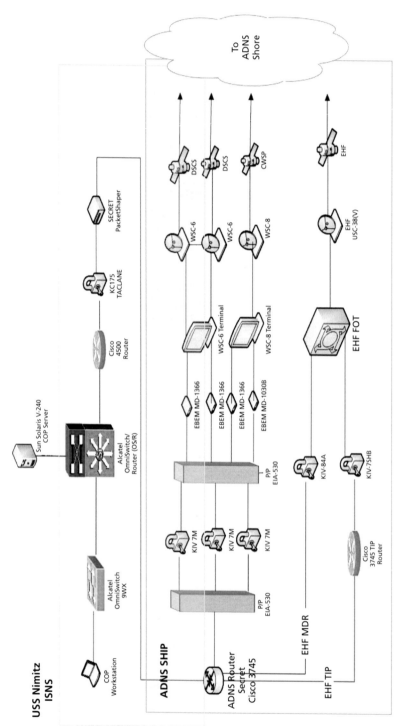

Figure C.2
Network Diagram of the COP Service (2 of 3)

ADNS Shore Architecture

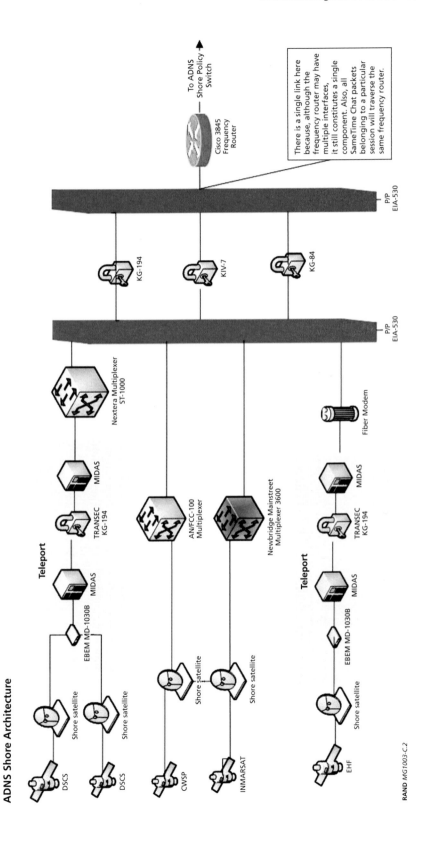

RAND *MG1003-C.2*

Figure C.3
Network Diagram of the COP Service (3 of 3)

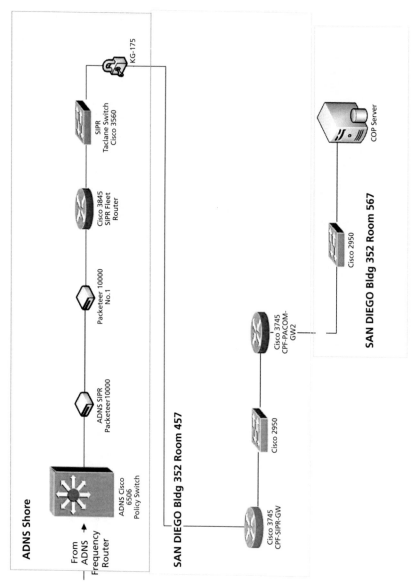

RAND *MG1003-C.3*

Bibliography

Bass, Tim, and Roy Mabry, "Enterprise Architecture Reference Models: A Shared Vision for Service-Oriented Architectures," draft submitted to MILCOM, March 17, 2004. As of January 15, 2009: http://www.enterprise-architecture.info/Images/Defence%20C4ISR/ enterprise_architecture_reference_models_v0_8.pdf

Birnbaum, Z. W., "On the Importance of Different Components in a Multicomponent System," in P. R. Krishnaiah, ed., *Multivariate Analysis II*, New York: Academic Press, 1969, pp. 581–592.

Boehm, Barry, *Software Engineering Economics*, New York: Prentice Hall, 1981.

Brandenburg, Craig, "U.S. Navy COTS: A Double-Edged Sword," briefing, National Defense Industrial Association Conference, Dallas, Tex., October 22–25, 2001. As of November 6, 2008: http://www.dtic.mil/ndia/2001systems/brandenburg2.pdf

Bryant, Gregory, "Interior Communications (IC) Way Ahead," briefing, Interior Communications Voice Symposium, Charleston, S.C., December 2, 2008. As of January 15, 2009: http://www.nicsymposium.com/Presentations/Tuesday/1_IC%20Way%20Ahead_Greg%20Bryant. pdf

Bryant, Gregory, and Allen Wolfe, "IC Voice Design and Acquisition: Comparison and Contrast," briefing, Interior Communications Voice Symposium, December 2–4, 2008.

Chairman of the Joint Chiefs of Staff, "Global Command and Control System Common Operational Picture Reporting Requirements," CJCS Instruction 3151.01B, October 31, 2008.

Coelho, Flávia Estélia Silva, Jacques Philippe Sauvé, Cláudia Jacy Barenco Abbas, and Luis Javier García Villalba, "Network Dependability: An Availability Measure in N-Tier Client/Server Architecture," in *Lecture Notes in Computer Science: Computer and Information Sciences—ISCIS 2003*, Vol. 2869, Springer Berlin/Heidelberg, 2003. As of April 30, 2010: http://www.springerlink.com/content/w5h6nc5vfwmvblyq/fulltext.pdf

Conwell, Candace, and Alan Kolackovsky, "CVN C4I CASREP Study," PEO C4I PMW 750 white paper, unpublished, and communication with authors, 2009.

Davis, Walter J., Jr., "Moving Forward," *CHIPS*, April 1995. As of April 30, 2010: http://www.chips.navy.mil/archives/95_apr/file1.html

Defense Daily, "Defense Watch," April 23, 2007. As of April 30, 2010: http://findarticles.com/p/articles/mi_6712/is_15_234/ai_n29347469/pg_1?tag=artBody;col1

EventHelix.com, "Reliability and Availability Basics," white paper, 2009. As of April 30, 2010: http://www.eventhelix.com/realtimemantra/FaultHandling/reliability_availability_basics.htm

Gartner, "Gartner Says Start Managing Information, Not Just Technology," press release, October 25, 2006. As of April 30, 2010:
http://www.gartner.com/it/page.jsp?id=546409

Gartner, *How Do You Measure Availability*, report G00150338, August 2007.

Gibson, Virginia Rae, and James A. Senn, "System Structure and Software Maintenance Performance," *Communications of the Association for Computing Machinery*, Vol. 32, No. 3, March 1989, pp. 347–358.

Gray, Jim, and Daniel P. Siewiorek, "High Availability Computer Systems," *IEEE Computer*, 1991. As of April 30, 2010:
http://citeseerx.ist.psu.edu/viewdoc/download;jsessionid=8269DB27EEB9649FAB8563089F6A04B0?doi=10.1.1.59.4466&rep=rep1&type=pdf

Hansen, Louis, "Navy Ship $840 Million Over Budget and Still Unfinished," *Virginian-Pilot*, June 30, 2007. As of April 30, 2010:
http://www.defense-aerospace.com/cgi-bin/client/modele.pl?prod=84137&session=dae.31670639.1197586699.4dYVZH8AAAEAAAMNJeUAAAAB&modele=feature

Henriksen, Steve, "Future Communications Study Technology Evaluation: SATCOM Availability Analysis," Working Paper, International Civil Aviation Organization, August 2006. As of April 30, 2010:
http://www.icao.int/anb/panels/acp/WG/M/Iridium_swg/IRD-05/IRD-SWG05-WP03%20-%20ITT%20SATCOM%20Availability%20Analysis.doc

Hong, John, "Network Consolidation Study Overview Briefing for Mr. Carl Siel, ASN(RDA) Chief Engineer," San Diego, Calif., August 29, 2007.

Hou, Wei, and O. G. Okogbaa, "Reliability Analysis for Integrated Networks with Unreliable Nodes and Software Failures in the Time Domain," *Proceedings of Annual Reliability and Maintainability Symposium*, 2000, pp. 113–117.

"Human Error Is the Primary Cause of UK Network Downtime; Intelliden Research Reveals that Lack of Standardization and Fat Fingers Cost Large Corporations and Governments Millions in Lost Revenue," M2 Presswire, November 6, 2003.

Information Technology Infrastructure Library, *Service Management*, London: Office of Government Commerce, 2007.

Jackson, William, "Carrier's Net Runs Aground," *Government Computer News*, November 19, 2001. As of April 30, 2010:
http://www.gcn.com/print/20_33/17535-1.html

Johnsson, M., B. Ohlman, B., A. Surtees, R. Hancock, P. Schoo, K. Ahmed, F. Pittman, R. Rembarz, and M. Brunner, "A Future-Proof Network Architecture," *Mobile and Wireless Communications Summit 2007*, July 2007.

Kiczales, Gregor, John Lamping, Anurag Mendhekar, Chris Maeda, Cristina Videira Lopes, Jean-Marc Loingtier, and John Irwin, "Aspect-Oriented Programming," *Proceedings of the European Conference on Object-Oriented Programming (ECOOP)*, Finland: Springer-Verlag, 1997.

Kuhn, D. Richard, "Sources of Failure in the Public Switched Telephone Network," *IEEE Computer*, Vol. 30, No. 4, April 1997, pp. 31–36.

Laprie, Jean-Claude, "Dependable Computing and Fault Tolerance: Basic Concepts and Terminology," in Proceedings of the 15th International IEEE Symposium on Fault Tolerant Computing (FTCS-15), Ann Arbor, Mich., June 1985, pp. 2–11.

Laprie, Jean-Claude, "Dependable Computing and Fault Tolerance: Basic Concepts and Terminology," in Jean-Claude Laprie, ed., *Dependability: Basic Concepts and Terminology*, Berlin: Springer-Verlag, 1992.

Larish, Brian, and Mike Ziegler, *End-to-End C4I Ao Pilot Study: Final Report (Initial Draft)*, Space and Naval Warfare Systems Command, August 22, 2008a.

———, "SPAWAR End-to-End C4I Ao Pilot Study," briefing, September 21, 2008b.

Lientz, B. P., and E. B. Swanson, "Characteristics of Application Software Maintenance," in *Communications of Association for Computing Machinery*, June 1978, pp. 466–471.

Malek, M., B. Milic, and N. Milanovic, "Analytical Availability Assessment of IT Services," in T. Nanya, F. Maruyama, A. Pataricza, et al., eds., "Service Availability—5th International Service Availability Symposium, Tokyo, Japan, May 19–21, 2008," *Lecture Notes in Computer Science*, Vol. 5017, 2008, pp. 207–224.

Matteo, Donald, Mark Rubin, William Clair, and Richard Bailer, "Program Manager Advisory Group (PMAG) Recommendations for the Integrated Communications & Advanced Network (ICAN)," PEO Aircraft Carriers, June 9, 2004.

Maxion, Roy, and Robert W. Reeder, "Improving User-Interface Dependability Through Mitigation of Human Error," *International Journal of Human-Computer Studies*, Vol. 63, Nos. 1–2, 2005, pp. 25–50.

McLaughlin, Laurianne, "Future-Proof Your Network," *Network World*, July 18, 2005.

Nagle, David, "ICAN Performs Well During Nimitz Exercise," press release, U.S. Navy, January 28, 2003a. As of April 30, 2010:
http://www.navy.mil/search/display.asp?story_id=5571

———, "Integrated Communications Network Performs Well in Nimitz Exercise," *Wavelength*, March 2003b.

National Research Council, *C4ISR for Future Naval Strike Groups*, Washington, D.C.: National Academies Press, 2006.

NAVAIR Lakehurst, NAVAIR Lakehurst fact sheet, 2002. As of April 30, 2010:
http://www.navair.navy.mil/lakehurst/NLWeb/documents/2002IEA.PDF

Naval Sea Systems Command Program Executive Office Ships Communications, "Navy Awards Contract for Additional LPD 17 Class Ships," press release, June 14, 2006. As of April 30, 2010:
http://aero-defense.ihs.com/news/2006/navy-lpd17-contract.htm?WBCMODE=presentationunpubli%2Cpresentationunpubli

O'Brien, Pat, "Applying US DoD Human Engineering Methods to Reduce Procedural Error Related Outages," *Proceedings of the 4th International Symposium on Service Availability*, Berlin: Springer, 2007, pp. 145–154.

O'Regan, Graham, "Introduction to Aspect-Oriented Programming," O'Reilly Media, January 14, 2004. As of April 30, 2010:
http://www.onjava.com/pub/a/onjava/2004/01/14/aop.html

Obert, Kathy, "Integrated Communications and Advanced Networks: Road Map," briefing, 1999.

Raytheon, "Raytheon's Shipboard Wide Area Network Ready for Navy's Next Generation Amphibious Transport Dock Ship, LPD 17," news release, April 21, 2003.

————, "Raytheon Delivers Source Code for DDG 1000 Total Ship Computing Environment Infrastructure Software," news release, November 7, 2006. As of April 30, 2010:
http://www.prnewswire.com/cgi-bin/stories.pl?ACCT=104&STORY=/
www/story/11-07-2006/0004468826&EDATE=

————, "Navy Approves Raytheon's Zumwalt Total Ship Computing Environmental Infrastructure," press release, October 30, 2007.

Reeder, Robert W., and Roy Maxion, "User Interface Dependability Through Goal-Error Prevention," International Conference on Dependable Systems and Networks (DSN'05), Yokohama, Japan, June 28–July 1, 2005.

Rosenberg, Barry, "Unified Approach: U.S. Navy Plans to Consolidate Networks," *C4ISR Journal*, June 1, 2008. As of April 30, 2010:
http://www.c4isrjournal.com/story.php?F=3502538

Schank, John F., Christopher G. Pernin, Mark V. Arena, Carter C. Price, and Susan K. Woodward, *Controlling the Cost of C4I Upgrades on Naval Ships*, Santa Monica, Calif.: RAND Corporation, MG-907-NAVY, 2009. As of April 30, 2010:
http://www.rand.org/pubs/monographs/MG907/

Schwartz, Mike, "USS NIMITZ (CVN 68) ICAN: INSURV INBRIEF," briefing, June 28, 2002.

Snaith, Pam, "The Changing Face of Network Management," CA Enterprise Systems Management white paper, October 2007. As of April 30, 2010:
http://www.ca.com/Files/WhitePapers/changing_face_network_mgmt_wp.pdf

Stark, George E., "Measurements to Manage Software Maintenance," Colorado Springs, Colo.: MITRE Corporation, 1997.

Stull, Mark, PEO C4I, briefing to the C4I executive committee, April 16, 2007.

Tokuno, Koichi, and Shigeri Yamada, "User-Perceived Software Availability Modeling with Reliability Growth," in T. Nanya, F. Maruyama, A. Pataricza, et al., eds., "Service Availability—5th International Service Availability Symposium, Tokyo, Japan, May 19–21, 2008," *Lecture Notes in Computer Science*, Vol. 5017, 2008, pp. 75–89.

Tolk, Andreas, and James Muguira, "The Levels of Conceptual Interoperability Model," paper presented at Fall Simulation Interoperability Workshop, Orlando, Fla., September 2003.

Tung, Teresa, and Jean Walrand, "Providing QoS for Real-Time Applications," University of California, Berkeley, 2003. As of April 30, 2010:
http://walrandpc.eecs.berkeley.edu/Papers/TeresaQoS.pdf

U.S. Department of Defense, *Human Engineering Guidelines for Military Systems, Equipment and Facilities*, MIL-HDBK-46855, 2006.

U.S. Navy, *Navy Training System Plan for the Aviation Data Management and Control System*, N78-NTSP-A-50-0009/A, March 2002. As of April 30 2010:
http://www.globalsecurity.org/military/library/policy/navy/ntsp/admacs-a_2002.pdf

Vellella, Frank, Ken East, and Bernard Price, "Reliability, Maintanability, and Availability Introduction," presentation, International Society of Logistics (SOLE), October 2008. As of April 30, 2010:
https://acc.dau.mil/GetAttachment.aspx?id=31987&pname=file&aid=5960&lang=en-US

Walsh, Edward J., "Navy's NIIN Aims for Single IT Shipboard Network Backbone," *Military and Aerospace Electronics*, November 2000.

Wang, Hao, Alex Gerber, Albert Greenberg, Jia Wang, and Yang Richard Yang, "Towards Quantification of IP Network Reliability," AT&T Labs, August 2007. As of May 17, 2010: http://www2.research.att.com/~jiawang/rmodel-poster.pdf

Wikipedia, "Differentiated Services," last updated December 5, 2008. As of April 30, 2010: http://en.wikipedia.org/wiki/Differentiated_services

Yamamura, Tetsuay, Kouji Yata, Tetsujiro Yasushi, and Haruo Yamaguchi, "A Basic Study on Human Error in Communication Network Operation," *Proceedings of the Global Telecommunications Conference*, 1989, and exhibition, *Communications Technology for the 1990s and Beyond*, IEEE, 1989.

Yau, S. S., and T. J. Tsai, "A Survey of Software Design Techniques," in *IEEE Transactions on Software Engineering*, June 1986, pp. 713–721.